SpringerBriefs in Molecular Science

Green Chemistry for Sustainability

Series editor

Sanjay K. Sharma, Jaipur, India

More information about this series at http://www.springer.com/series/10045

Kalyani Barve · Apurva Dighe

The Chemistry
and Applications
of Sustainable Natural
Hair Products

Kalyani Barve
Shobhaben Pratapbhai Patel School
 of Pharmacy and Technology
 Management
Mumbai
India

Apurva Dighe
Shobhaben Pratapbhai Patel School
 of Pharmacy and Technology
 Management
Mumbai
India

ISSN 2191-5407   ISSN 2191-5415 (electronic)
SpringerBriefs in Molecular Science
ISSN 2212-9898
SpringerBriefs in Green Chemistry for Sustainability
ISBN 978-3-319-29417-9   ISBN 978-3-319-29419-3   (eBook)
DOI 10.1007/978-3-319-29419-3

Library of Congress Control Number: 2016930175

Printed on acid-free paper

This Springer imprint is published by SpringerNature
The registered company is Springer International Publishing AG Switzerland

Preface

The book has been written to provide reference material for those involved in the formulation and development of hair products, namely hair oils, shampoos, conditioners and hair dyes. With the advent of synthetic ingredients being added to hair cosmetics, the incidences of hair damage have increased simultaneously. This can be curbed only with the use of natural products. These have phytoconstituents as their main active ingredients which are nothing, but chemicals. The holistic use of these natural products is the one that makes the difference. Since the whole plant or a fraction of it is being used, it has a number of chemical entities present, some of which synergise the efficacy while others mar the side effects.

The authors are thankful to editors, Dr. Sonia Ojo and Dr. Esther Rentmeester, Mr. Ravi Vengadachalam, project co-ordinator, and Springer for giving an opportunity to publish with Springer. The authors would also like to acknowledge the support provided by Shobhaben Pratapbhai Patel School of Pharmacy and Technology Management, SVKM's NMIMS in creating this book.

Contents

Chapter 1
Hair

Abstract This chapter gives an overview of the anatomy and functioning of hair. It also provides information on the importance of hair as fashion statement. To modify ones looks, people try different cosmetics not just on the skin but also for hair. Most of the hair cosmetics available today have all harmful synthetic chemicals into them. So there's a need to look out for safe cosmetics which could be provided only from natural sources like plants.

Keywords Keratin · Hair cosmetics · Natural products · Plants

1.1 Introduction

Hair is one of the vital parts of body and a describing characteristic of mammals. It has two features:

1. Hair follicle which lies beneath the skin which helps in re growth of hair and skin as it has the stem cells. Tens or thousands of them are deeply invaginated in the scalp. The dermal papilla is present at the base of the hair follicle, which receives blood supply and provides nourishment to produce new hair. It also contains receptors for testosterones, which regulate hair growth.
2. The filamentous part above the skin which is called a shaft. It is formed as a process of protein synthesis, structural alignment and completion of the keratinization of the follicular cells. The shaft has three layers:
   (a) The outermost layer called the cuticle having flat thin cells.
   (b) The middle layer containing the fibrillary structure of keratin and pigment melanin. This gives strength, colour and texture to hair.
   (c) The innermost layer known as the medulla is a centrally vacuolated area.

In addition to this the hair also supports the sebaceous and sweat glands. With this structure hair functions as a protective appendage. It gives a feeling of warmth and also helps in the sensation of touch (Krause and Foitzik 2006; Feughelman 1997).

© The Author(s) 2016   1
K. Barve and A. Dighe, *The Chemistry and Applications of Sustainable Natural Hair Products*, SpringerBriefs in Green Chemistry for Sustainability, DOI 10.1007/978-3-319-29419-3_1

Around the second month of pregnancy, hair follicles begin to develop in the embryo. The major embryological stages of hair are

1. **Pre-germ stage**: Mesenchymal cells aggregate in the dermis and the basal epidermal cells start thickening simultaneously.
2. **Germ stage**: There is an elongation of the basal epidermal cells to form the hair germ and simultaneous replication of mesenchymal cells to form the dermal papilla.
3. **Hair-peg stage**: Growth of hair germ cells downwards and formation of hair peg, which propel the mesenchymal cells further downwards.
4. **Bulbous-peg stage**: The hair peg forms the apocrine glands and sebaceous glands. The arrector pili develop near the sebaceous gland and attach itself to the hair peg.
5. **The first primordial hair**: Proliferation of the cells around the dermal papilla and formation and emergence of hair shaft.

Normal hair growth cycle
The approximate growth of hair is 10 cm per year. Hair follicles undergo three phases of growth in a cyclic manner.

1. *Anagen*—It is also known as the growth phase which may vary from 2–8 years.
2. *Catagen*—It is also known as the transitional phase which lasts for 1–2 weeks. At the end of the growth phase the hairs enters into a this phase. There happens shrinkage of hair follicles, breaking of the dermal papilla to rest below, during this phase.
3. *Telogen*—It is also known as the resting phase, which lasts for 5–6 weeks. This phase comes after the transitional phase. As the name indicates the hair follicle along with the dermal papilla remain in a resting phase. Both of them join and a new hair growth begins at the end of this stage. This new hair pushes the old one out and hair follicle enters in the growth phase again.
   Each hair passes through all these phases independent of the neighbouring hairs (Ralf and George 1999; Bertolino et al. 2003).

Differences in hair
Every individual will have different types of hair. It may vary in the thickness, shape, colour or texture. Hair colour is due to the presence of two pigments namely: Eumelanin which is a dark pigment and gives brown or black hair and Pheomelanin which is a light pigment and gives blonde or red shade to hair.

Chemical composition of hair
Hair is essentially a cross linked, partially crystalline, oriented polymeric network which contains a number of functional chemical groups. Proteins are the primary components of hair, keratin being the principal protein. The mineral content of human hair is from 0.25 to 0.95 %. Human hair contains large amount of amino acids like glycine, threonine, aspartic and glutamic acid, lysine, cysteine and tyrosine, trace elements (secretary glands), sebum-free fatty acids, combined fatty

acids, and like squalene, cholesterol, and waxes (sebaceous glands), sweat-water salts such as Na+ and K+ along with urea, amino acids, lactate and pyruvate (sweat glands) (Kapoor 2005).

Need for hair care products

Humans since ancient times always wanted to be attractive, which could be achieved by modifying the external appearance. Skin, nails and hair are the most common targets for cosmetic modifications to enhance ones charm. Hair cosmetics mainly focus on hair growth, hair types, hair colour and hair care which play an important role in the physical appearance. To modify the hair cosmetically, alterations need to be done in the cortex, medulla or the cuticle. Hair can be modified both externally as well as internally with the use of different cosmetics.

Most of the hair care cosmetics have numerous chemicals in its formulation. Synthetic and chemical based products cause human health hazards and have several side effects. Few examples are quoted below.

Lead and Paraphenylenediamine found in hair dyes may cause serious effects on the central nervous system, mild dermatitis or affect vision. Formaldehyde used in the shampoo and bleach may harm the pulmonary system and elevated exposure may cause hypersensitivity. Dibutylpthalates incorporated as fragrance for shampoo may cause gastrointenstinal disturbance in humans. Surfactants like diethanolamine, monoethanolamine, triethanolamine, lauramide and propylene glycol used in styling gels/lotions, conditioners, shampoos, hair dyes may cause irritation of the mucous membranes. Hydroquinone used in hair bleach may cause tinnitus, nausea, shortness of breath, cyanosis and convulsions.

One way of reducing or eliminating these side effects is minimizing the use of such chemicals and replacing them with natural products. Personal hair care industry has now shifted to the use of natural herbal care products along with the bioactive constituents from various botanicals which would serve as a cosmetic for hair and provide nutrients. Plants provide vitamins, antioxidants, essential oils, proteins, hydrocolloids which can be used in the formulations. Ancient medicinal systems in India such as *Ayurveda*, *Siddha* and *Unani* could be of great help to identify the phytochemicals which can act as a wonders for the hair care products.

References

Bertolino AP, Klein LM, Fredberge LM (2003) Biology of hair follicles. In: Fitzpatrick TB, Eisen AZ, Wolf K, Fredberg IM, Austen KF (eds) Dermatology in general medicine. McGraw Hill Inc, New York

Feughelman M (1997) Mechanical properties and structure of alpha-keratin fibres: wool, human hair and related fibres. UNSW Press

Kapoor VP (2005) Herbal cosmetics for skin and hair care. Nat Prod Radiance 4(4):306–314

Krause K, Foitzik K (2006) Biology of the hair follicle: the basics. Semin Cutan Med Surg 25:2–10

Ralf P, George C (1999) The biology of hair follicles. New Engl J Med 341:491–497

Chapter 2
Hair Oils

Abstract We are exposed to lots of chemicals from our environment either knowingly or unknowingly. These chemicals enter our body and induce damage to different organs. One such organ affected is the hair. Damage to hair, weakens them and causes hair fall. In order to reverse this damage, the nourishment to hair has to be reinstated which is possible by increasing blood circulation to the scalp and also by massaging with oils that provide external nutrients. This chapter provides a list of plant drugs that can be incorporated in hair oils, along with their beneficial effects for hair.

Keywords Aloe · Green tea · Coconut oil · Hibiscus

2.1 Introduction

Hair oils come under the category of hair tonics used for hair disorders such as dandruff, hair fall and dryness of hair. Regular application of hair oil provides resistant to breakage with no split ends, provides luster and shine to hair and moisturizes the scalp. Hair oils also act as a medium to supply essential nutrients to the root of hair for their proper growth. Below we have mentioned few ingredients which can be incorporated in hair oils for the above mentioned uses.

The base oils has to be a non-sticky oil and the choice may be one of the following.

2.2 Types of Hair Oils

2.2.1 Almond Oil

It is the oil obtained from the fruits of *Prunus dulcis/Prunus amygdalus*, family Rosaceae. It is native to Middle East and South Asia. The oil is extracted by

K. Barve and A. Dighe, *The Chemistry and Applications of Sustainable Natural Hair Products*, SpringerBriefs in Green Chemistry for Sustainability, DOI 10.1007/978-3-319-29419-3_2

pressing the dried fruits either by cold or hot press. The other alternative for almond oil extraction is solvent extraction (Li et al. 2006). The crude oil is purified by filtration. In *Ayurveda* almond oil has been attributed with aphrodisiac properties and as a nervine nutrient (Burlando et al. 2010).

Chemical composition
Almond oil comprises of monounsaturated fatty acid oleic acid and polyunsaturated fatty acid i.e. linoleic acid. The others include stearic acid, palmitic acid and linolenic acid. It is also a rich source of Vitamin E.

Benefits for hair
It is used to soften and moisturize dry hair. It is also used to stimulate hair growth and provide elasticity to the strands making them strong. It also helps to cleanse the scalp (Kumar et al. 2012).

2.2.2 Argan Oil

The oil is obtained from the kernels of *Argania spinosa*, family Sapotaceae, found in the desert regions of Morocco. The fruit is dried, pulp removed and the nut is opened to get the kernel. These kernels are roasted, cooled and then grinded to get the oil. The oil is allowed to stand, it is then decanted and filtered to remove the impurities.

Chemical composition
The oil contains tocopherols, phenols, major fatty acids like oleic, palmitic, stearic, linoleic and linolenic acids (Rueda et al. 2014). It also contains squalene, steroids-schottenol and spinasterol (Khallouki et al. 2003).

Benefits for hair
The oil is considered to be nourishing for hair and also a good moisturizer. It brings shine to hair (Monfalouti et al. 2010). It also prevents hair loss (Guillaume and Charrouf 2011).

2.2.3 Apricot Seed Oil

It is extracted from the kernels of *Prunus armeniaca*, family Rosaceae. It is grown wildly in the northern part of India, USA, Australia. It is indigenous to East and Central Asia, largest producer being Turkey. The pulp of the fruit is separated, followed by separation of the kernels from the seed coat. The kernels are then put into expellers to get the oil. The oil obtained using this method may have the presence of hydrocyanic acid. An alternative method is solvent extraction, which eliminates the presence of hydrocyanic acid in the oil (Gupta and Sharma 2009).

Chemical composition

The oil is very similar to almond oil. The oil contains unsaturated fatty acids namely palmitic, palmitoleic, linoleic, linolenic and oleic acid. The oil also shows the presence of proteins (Dwivedi and Ram 2008). It also contains small quantity of glycolipids, phospholipids, sterols, tocopherols, amino acids and minerals (Alpaslan and Hayta 2006).

Benefits for hair

It provides a conditioning effect to the hair. It keeps the scalp moisturized, stimulates the hair follicles and enhances hair growth, due to presence of Vitamin E. It makes the hair soft and manageable.

2.2.4   Babassu Oil

It is extracted from the kernels of the palm tree *Attalea speciose* (*Orbignya oleifera*), belonging to the family Arecaceae. It is found in the Amazon region. Once the fruits are ripe the nuts are opened and the oil is cold pressed from the kernels.

Chemical composition

The oil has similar composition as that of coconut oil and comprises of lauric, myristic, palmitic, oleic and stearic acid (Ferrari and Soler 2015).

Benefits for hair

One advantage of this oil over coconut oil is it doesn't clog the pores of the scalp. It may be used as a pre shampoo treatment. It can also be mixed with shea butter or other ingredients to enhance its efficacy. It provides excellent moisturizing effect and can soothe itchy and dry scalp. There are many hair care formulations that make use of babassu oil.

2.2.5   Castor Oil

The oil is obtained from the seeds of *Ricinus communis*, belonging to family Euphorbiaceae. The plant is found in all tropical and sub-tropical countries, India being the major producer. The oil is obtained from the decorticated seeds of the plant which are fed in the oil press to get the oil. This oil is filtered and steamed to remove ricin, a toxic component in the oil and lipase to prevent rancidity in the oil.

Chemical composition

It chiefly contains glycerides of ricinoleic, isoricinoleic, stearic and dihydroxy-stearic acid. Ricinoleic acid an omega-9 fatty acid is the main component of castor oil, is used for the treatment of baldness.

Benefits for hair
Hydrogenated castor oil which is a white waxy solid is also used in different hair cosmetics. It is extensively used as a hair tonic. It is also incorporated in many hair care preparations (Sagarin 2008). Being rich in omega-9 fatty acids, it provides nourishment to the hair and follicle. It has a good penetrability and hence can easily provide nourishment to the follicles.

2.2.6 Coconut Oil

It is the oil obtained from the kernel or meat of fruits of *Cocus nucifera*, family Arecaceae. It is indigenous to India, Sri lanka, Indonesia, Malaysia, Australia and South America. It can be used for different purposes like deep conditioning, to control dandruff, for hair growth and as a base for hair colour. For the preparation of oil, there are two processes used:

(a) Dry process which makes use of the dried pulp of coconut. It is pressed to obtain crude oil. The crude oil is then refined to remove impurities.
(b) Wet process which extracts the oil from the raw coconut. The oil obtained is initially in the form of an emulsion due to the presence of proteins. The oil is extracted by breaking the emulsion either by boiling, centrifugation or pre-treatment with salts, enzymes or acids (McGLONE et al. 1986).

Chemical composition
Coconut oil contains saturated fatty acids and is one of the richest source of medium chain fatty acids. The major fatty acids are lauric acid, myristic acid and capric acid. (Marina et al. 2009). It additionally contains phenolic acids and antioxidants like tocopherol (Appaiah et al. 2014).

Benefits for hair
Coconut oil prevents the protein loss from hair (Rele and Mohile 2003). It is generally used as a pre-wash conditioner. It also prevents the penetration of water in the hair shaft and the subsequent swelling and chipping off the hair cuticle. It also prevents the loss of moisture from hair and thus provides moisturizing effect (Keis et al. 2007). It helps to control dandruff (Fife 2013).

2.2.7 Hemp Oil

The oil is extracted from the seeds of the plant *Cannabis sativa*, family Cannabaceae. It is common to be confused with the banned entity marijuana and hemp plant. The oil is obtained from industrial hemp plants, which contain very low levels of tetrahydrocannabinol (THC) (Bosy and Cole 2000). The oil is extracted by cold press method, solvent extraction (petroleum ether or naptha) or by more recent technique like supercritical CO_2 extraction.

Chemical composition
The oil is rich in essential fatty acids, proteins—edestin and albumin, amino acid—arginine and vitamins (Callaway 2004). It contains high levels of linoleic acid along with α- and γ-linolenic, palmitic, stearic, oleic acids. In addition it also contains tocopherols (Anwar et al. 2006).

Benefits for hair
Hemp oil moisturizes the scalp and hair, stimulates hair growth and prevents hair breakage. Massaging with hemp oil increases blood circulation to the scalp. It also helps to control dandruff.

2.2.8   Jojoba Oil

It is the liquid present in the seeds of *Simmondsia chinensis*, family Simmondsiacaceae. It is native to South America: Arizona, California and Mexico. The oil is obtained by pressing the seeds followed by purification using solvents for leaching the oil. An alternative for extraction of oil is direct solvent extraction.

Chemical composition
It contains wax esters of fatty acids, major ones being eicos-11enoic, octoadec-9-enoic docos-13 enoic, the major alcohol being docos-13-enol (Wisniak 1987).

Benefits for hair
The pure oil is used as a moisturizer for hair. The liquid wax has high affinity for sebum which allows high penetration and delivery to deeper skin layers. It provides hydration, antibacterial and anti-parasitic effects. It is used clinically to strengthen hair. It is used to cleanse the scalp and hair. It reduces the oiliness at the same time improving gloss and manageability of hair (Wisniak 1987).

2.2.9   Linseed Oil

It is the oil obtained from the seeds of *Linum usitatissimum*, Linaceae. It is indigenous to India, but also cultivated in Canada, China and US, Turkey and Afghanistan. For the extraction of oil, the seeds are made free of other materials by sieving and oil is produced by the use of expellers. The oil obtained this way is subjected to decanting, alkali treatment followed by bleaching, to remove the mucilage, free fatty acids and coloring matter respectively. The oil is used as a nutraceutical and is recommended to prevent coronary heart disease, cancer, neurological and hormonal disorders.

Chemical composition
The oil contains glycerides of palmitic, stearic, oleic, linoleic and linolenic acids. It also contains sterols, tocopherol and squalene.

Benefits for hair
As a hair oil it helps to control dandruff, alleviate hair fall and also works as a conditioner for dull hair (Tripathi et al. 2013).

2.2.10 Mustard Oil

Commonly referred as Sarson ka tel in India, it is obtained from the seeds of *Brassica nigra,* family Cruciferae. The plant is cultivated in India, china, Canada and England. The oil is obtained by expression of the seeds, followed by purification.

Chemical composition
It contains glycerides of arachidic, behenic, eicosenoic, erucic, lignoceric, linoleic, linolenic, oleic, palmitic and myristic acids (Chowdhury et al. 2007). Apart from these, it contains a glycoside sinigrin, tocopherol and carotenoids (Vaidya and Cho 2011).

Benefits for hair
It is an edible oil but also used for its medicinal properties such as a rubefacient. It is used for hair growth promoting effects, to prevent premature greying of hair and hair loss. It may also help to control infections of the scalp. It improves the texture of hair and provides additional conditioning effect.

2.2.11 Nigella sativa *Oil*

It is commonly known as black cumin or kalonji. The oil is obtained from the fruits of *Nigella sativa*, family Ranunculaceae. It is indigenous to Europe, Southwest Asia, Africa but is cultivated in other countries. The oil may be extracted by cold press method or solvent extraction.

Chemical composition
The oil contains unsaturated fatty acids: linoleic acid, oleic acid, eicosadienoic acid and dihomolinoleic acid, saturated fatty acids like palmitic and stearic acid. Thymoquinone, nigellone, α-sitosterol, and stigmasterol as major components (Ahmad et al. 2013).

Benefits for hair
It has been used in the Unani system of medicine for treatment of hair problems. The oil strengthens the follicles and prevents hair loss. It also claims to treat baldness by causing hair re-growth.

2.2.12 Olive Oil

It is the expressed from the pericarp of fruits of *Olea europoea*. Family Oleaceae. It is a native to Mediterranean countries. The ripe fruits are collected, the pulp removed and the seeds are pressed lightly to extract the oil. This oil is mixed with water to remove coloring impurities and oil is skimmed off.

Chemical composition
The oil chiefly contains olein and a little of palmitin, linolein and arachin (Wallis 2005, Textbook of Pharmacognosy). The virgin olive oil contains many antioxidants like phenolics, flavonoids, lignans and secoiridoids (Servili et al. 2009).

Benefits for hair
It has emollient properties. It forms a protective coating over the hair shaft and prevents from damage. It may be used for the treatment of dandruff, to moisturize and soften hair.

2.2.13 Rice Bran Oil

The rice bran comprises of the embryo and endosperm of the seeds of *Oryza sativa*, family Graminae. It is produced in Asian countries like China, India, Bangladesh and Japan. The fresh bran is pressed and then extracted with solvent (hexane) to get the oil. The oil can also be obtained by supercritical CO_2 extraction.

Chemical composition
It contains glycerides of oleic, linoleic and palmitic acid and antioxidant like tocopherols. It also contains oryzanol which is a mixture of ferulic acid esters of triterpene alcohol, one of the active constituents (Patel and Naik 2004). It contains oleic, linoleic, alpha-linolenic, palmitic and stearic acid. It also contains free sterols (Cicero and Gaddi 2001).

Benefits for hair
It stimulates hair growth and makes hair thicker. It strengthens the hair follicles. It is used as a hot oil treatment. It may help to control dandruff by moisturizing the scalp. It also acts as a sunscreen (Nagendra Prasad et al. 2011) and also prevents split ends. Linoleic acid and γ-oryzanol induce hair growth by stimulating the

mRNA expression levels of VEGF (vascular endothelial growth factor), IGF-1 (insulin like growth factor-1), KGF (keratinocyte growth factor) and reduction of TGF-beta (transforming growth factor beta) and thus could be used for treatment of hair loss (Choi et al. 2014). It may also help to prevent greying of hair (Unna and Sampson 1940).

2.2.14 Rosehip Oil

It is the oil obtained from the rose hip seeds, *Rosa canina, R. moshata, R. rubiginosa* belonging to family Rosaceae. The oil is extracted either by cold press method or solvent extraction method from plants growing in the Andes, South Africa and Europe. Recently the supercritical CO_2 extraction is being advocated.

Chemical composition
The major fatty acids found in the oil are linolenic, palmitic and stearic acid (Szentmihályi et al. 2002). It has other nutrients like tretinoin, lycopene and beta carotene.

Benefits for hair
It works as a good moisturizer and is good for dull and dry hair. It is also said to cure dandruff and provide shine to hair. One of the study indicates that the incorporation of oil prevents damage to hair following the use of permanent wave solution (Miyeon and Kyoungsook 2013). Since the oil is rich in antioxidants it may help to prevent the damage to hair follicles and prevent graying of hair.

2.2.15 Safflower Oil

It is obtained from the ripe seeds of *Carthamus tinctorius*, family Compositae. It is indigenous to Middle East but is largely grown in India, USA, Ethiopia and Mexico (Ekin 2005). The oil is prepared by expression with a hydraulic press. The oil is filtered and decolorized.

Chemical composition
It contains glycerides of palmitic, stearic, arachidic, oleic alinoleic and linolenic acid. It also shows the presence of tocopherols and tocotrienols (Lee et al. 2004).

Benefits for hair
The oil is medicinally used as a dietary supplement in the treatment of hypercolesteremia and atherosclerosis. The advantages of using safflower oil is it is inexpensive. It provides excellent moisturizing effect. It is used for hair growth, to impart shine to hair and to prevent hair loss.

2.2.16   Sea Buckthorn Oil

The oil is obtained from the seeds of the berries of *Hippophae rhamnoides*, Elaeagnaceae family. It has long been used by the Tibetians for various health problems. It is native to the sandy areas along the coast of Europe and Asia. The oil may be extracted by any one of the following methods: cold or hot press, solvent extraction, maceration in other carrier oils, but the best suited is supercritical CO_2 extraction.

Chemical composition
It is a rich source of omega 3, 6, 7 and 9 fatty acids and vitamins essential for hair growth and good health of hair. The oil is rich source of tocotrienols, tocopherols, carotenoids, vitamin A, C and E and trace elements. The major fatty acids are linoleic, alpha-linoleic, palmitic and palmitoleic acid (Suryakumar and Gupta 2011).

Benefits for hair
The oil needs to be mixed with a carrier oil like coconut, jojoba or olive oil and massaged onto the scalp. Regular use of sea buckthorn oil prevents premature hair fall, helps to clean the hair follicles, aids in hair elasticity and moisturizes the hair shaft. It has been used in Russia for treatment of dandruff and prevention of hair loss (Ruan et al. 2007).

2.2.17   Sesame Oil

The plant being indigenous to India, is commonly known as til oil in India. It is obtained from the seeds of *Sesamum indicum*, family Pedaliaceae. It is found in India, China, Egypt and Middle East. The white variety of seeds are expressed to yield oil which is then purified by refining methods.

Chemical composition
It contains glycerides of oleic, linoleic, palmitic, stearic and arachidic acids. It also contains sesamol responsible for the stability of oil and lignin derivatives: sesamin and sesamolin.

Benefits for hair
It has very good emollient properties and is used as an oil base for many *Ayurvedic* hair oils (Angadi 2009). Regular use of sesame oil helps to have healthy hair. It also enhances the growth of hair and prevents premature greying of hair.

2.2.18   Sunflower Oil

It is obtained after the cold compression of seeds of *Helianthus annuus, belonging to family* Asteraceae. It is the second most widely used base oil in the hair cosmetics.

Chemical composition
It contains mono and polyunsaturated fatty acids mostly oleic and linoleic acids and minor quantities of palmitic and stearic acid (Putt and Carson 1969). It also contains carotenoids and tocopherols.

Benefits for hair
It stimulates the natural growth of hair, acts as a good moisturizer and conditioner. It prevents thinning of hair, nourishes the hair and prevents damage to hair.

2.2.19 Wheat Germ Oil

It is extracted from the germ of kernels of wheat, *Triticum aestivum*, family Graminae. It is largely cultivated in Asia and Europe. The oil is extracted via pressing or solvent extraction (Knowles et al. 2014).

Chemical composition
The oil consists of fatty acids: palmitic acid, oleic, linoleic and linolenic acid, (Megahad and El Kinawy 2002) tocopherols, sterols, flavonoid pigments and xanthophylls (Barnes 1982). Wheat germ oil contains vitamin E, A and D which is easily absorbed by the skin and the scalp. Also at the same time it prevents the oil from getting rancid. It also contains proteins and lecithin (Kumar et al. 2011).

Benefits for hair
The oil may be used for the repair of dry and dull hair. It may be possible to delay hair loss if not complete prevention using this oil. It stimulates hair growth and helps to maintain the elasticity of hair.

    Below we have mentioned few ingredients which can be incorporated in hair oils to enhance benefits of the oils.

2.3 Herbal Oil Ingredients

2.3.1 Allium cepa *(Onion)*

It is commonly known as onion, belongs to the family Liliaceae. It is the most widely cultivated species in this genus. The crop is harvested after the leaves wither away and dried onions are ready for use or storage.

Chemical constituents
The seeds show high amounts of fixed oil, fibre, crude protein, calcium, potassium, low amounts of sodium and cysteine derivatives. It is a rich source of free amino

acids such as Threonine, Aspartic acid, Valine, Asparagine, Glutamic acid, Serine, Glycine, Glutamine, Histidine, Arginine, Alanine, Proline, Tyrosine and GABA (Dini et al. 2008).

Benefits to hair
Sulfur is the building block of hair, onions being rich source of sulfur help in providing the required nourishment and helps in hair growth. It prevents the premature graying of hair and also have beneficial effects for scalp infections due to anti-bacterial property. A blend of three tablespoons of onion juice, one tablespoon of coconut oil and a tablespoon of olive oil, massaged into the scalp at least three times a week, is good for treating hair loss, particularly if the scalp feels dry and itchy. A blend of onion juice and honey when massaged on the scalp, provides nourishment to hair.

2.3.2   Allium sativum *(Garlic)*

It is commonly known as garlic, is native to central Asia and has long been a staple as well as a frequent seasoning in the Mediterranean region.

Chemical constituents
It contains carbohydrates and proteins. The major sulfur-containing compounds in intact garlic are γ-glutamyl-*S*-allyl-L-cysteines and *S*-allyl-L-cysteine sulfoxides. It also contains a sulfur containing amino acid alliin which is converted to allicin which is a thiosulfinate when garlic is cut or crushed. *S*-allyl-cysteines (SAC), a major transformed product from γ-glutamyl-*S*-allyl-L-cysteine, is a sulfur amino acid. The volatile components include diallyl sulfide (DAS), diallyl disulfide (DADS), diallyl trisulfide, methylallyl disulfide being the major ones. It also contains phosphorous, iron and copper (Amagase 2006).

Benefits to hair
The benefits of garlic are again attributed to the presence of sulfur and would be the same as mentioned for onion. Garlic oil may be incorporated in the hair oil, but to mask the odour of garlic another strong aromatic constituent needs to be added. Lemon oil can be on alternative to mask the strong odor of sulfides.

2.3.3   Aloe barbadensis *Mill* (Aloe vera)

Aloe vera is a succulent plant species. It is indigenous to eastern and southern Africa and cultivated in Europe and many parts of India. The species is frequently cited as being used in herbal medicine. Extracts from *Aloe vera* are widely used in the cosmetics and alternative medicines, claiming to have rejuvenating, healing, or soothing properties.

Chemical constituents

The principle component is aloin, which is a mixture of glucosides. Amongst these glucosides, barbaloin is the chief constituent. Barbaloin is a c-glycoside having aloe-emodin anthrone as its aglycone. The other constituents include isobarbaloin, β-barbaloin, aloe-emodin and resins. The drug also contains aloetic acid, homonataloin, aloesone, chrysophanic acid, chrysamminic acid, galactouronic acid, choline, choline salicylate, mucopolysachharides, glucosamines, hexuronic acid and coniferyl alcohol. It also contains sugars such as arabinose, galactose, glucose, mannose, rhamnose, xylose, hexuronic acids and sterols such as cholesterol, campesterol, sitosterol and lupeol (Waller et al. 1978).

Three new chromones has been isolated which are 5-(hydroxymethyl)-7-methoxy-2-methylchromone, 5-((4E)-2′-oxo-pentenyl)-2-hydroxymethylchromone, and 7-hydroxy-5-(hydroxymethyl)-2-methylchromone (Zhong et al. 1991).

Benefits to hair

Aloe contains salicylate, carboxypeptidase and magnesium lactate, all being anti-inflammatory in nature. This effect is particularly of benefit in inflammatory conditions like seborrheic dermatitis. Mucopolysaccharides found in aloe gel, are reported to heal wound and could be used to heal scalp wounds. They also provide moisturizing effect to the scalp, improve blood circulation and promote the growth of hair. Thus aloe gel may be used directly as a replacement for hair oil. Certain other ingredients can also be added to aloe gel to improve its efficacy in treating different hair problems (Waller et al. 1978).

2.3.4   Camellia sinensis *(Green Tea)*

Green tea is indigenous to China, is made from the leaves from *Camellia sinensis* that have undergone minimal oxidation during processing.

Chemical constituents (Graham 1992)

Fresh tea leaf is rich source of polyphenols mainly catechins. Other polyphenolic compounds include flavonoids, chlorogenic acid, coumarylquinic acid and theogallin and theanin, both being unique to tea. The content of caffeine in green tea is very les along with other alkaloids like theobromine and theophylline.

Benefits to hair

The chief polyphenol in green tea is epigallocatechin gallate (EGCG) which promotes hair growth in cell culture. It activates the hair follicles which promote hair growth. It can destroy the bacterial and fungal parasites, which weaken the roots of hair and causes hair fall. It is also known that the methyl xanthine type of alkaloids, caffeine being one of them is used for the treatment of androgenic alopecia.

2.3.5 Capsicum minimum

It is also known as cayenne pepper or red pepper, named after the city of Cayenne in French Guinea. The fruits are generally dried and ground, or pulped and baked into cakes, which are then ground and sifted to make the powdered spice of the same name.

Chemical constituents
The main constituent of pepper is capsaicin, which is the pungent phenolic fraction of capsicum. It also contains 6,7-dihydrocapsaicin, ascorbic acid, thiamine and fixed oil. It has several minerals like calcium, phosphorus, iron and potassium and vitamins like thiamine, riboflavin and niacin. *C. annuum* is rich in carotenoid pigments which include capsanthin, capsorubrin, carotene, luteine, zeaxanthin and cucurbitaxanthin A. It is also rich in fats, proteins, Vitamin A and vitamin C.

Benefits to hair
Capsaicin being pungent in nature casuses a tingling sensation when applied to the scalp which activates the hair follicles. In addition it increases blood circulation, this enhances blood supply to the follicles provides the required nutrients for hair growth. The quantity of the extract to be used in the oil should be sufficient so as to tingle the scalp but not too much so as to cause irritation of the scalp.

2.3.6 Centella asiatica

It is the whole herb of *Centella asiactica*, belonging to family Umbelliferae, found in wet areas in Insia, Sri Lanka, Pakistan, Indonesia, Africa, Australia. It shows sedative, spasmolytic, anti-anxiety action. It is also used to cure skin diseases.

Chemical constituents
It mainly contains saponins called asiactcoside and madecassoside which yield triterpene acids Asiatic acid and madecassic acid on hydrolysis.

2.3.7 Cocos nucifera *(Coconut Milk)*

It known to the world as coconut and belongs to the family Arecaceae, being indigenous to the tropical and subtropical regions. The inner part of the fruit has lots of water and the endosperm is suspended in it. Gradually the endosperm starts depositing layers along the wall of coconut and forms the flesh, which is an edible part. This flesh when made into a paste and filtered yields coconut milk. The same flesh when dried also yields coconut oil.

Chemical constituents
It is rich source of inorganic ions such as calcium, iron, magnesium, phosphorous, potassium, sodium, copper, zinc, chlorine, sulphur, vitamins like vitamin C, B1, B2, B3, B5, B6, Folic acid, Biotin and Nicotinic acid and amino acids like Alanine, β-Alanine, γ-Aminobutyric acid, Arginine, Asparagine and glutamine, Aspartic acid, Cystine, Glutamic acid, Glutamine, Glycine, Homoserine, Histidine, Isoleucine, Leucine and Phenylalanine (Yong et al. 2009).

Benefits to hair
Coconut milk is rich in iron, potassium and essential fats. It reduces hair fall and breakage. It also promotes hair growth and thus could be one of the essential components in hair oil. Coconut oil derived from the dried flesh itself has long been used as a base for hair oils and has reported benefits for growth of hair.

2.3.8 Cucurbita pepo

It is an oil obtained from the seeds of pumpkin, family Cucurbitaceae. It is a European union protected designation of origin (PDO) product and is an important export commodity of Austria and Slovenia.

Chemical constituents
The oil contains fatty acids, predominantly: linoleic, oleic, palmitic, and stearic acids. It also contains a higher quantity of tocopherol (α and β) and carotenoids (Stevenson et al. 2007; Procida et al. 2013).

Benefits to hair
A recent study has indicated the usefulness of pumpkin seed oil in the treatment of androgenetic alopecia (AGA), which is probably because of the presence of phytosterols which inhibit 5α-reductase. There are few more plants that do have an effect especially on AGA. *Cuscuta reflexa* via 5α-reductase inhibitory activity, *Panax ginseng* due to anti-inflammatory and blood circulation effect, *Eclipta alba* extract and *Zizyphus jujuba* essential oil, all have showed possibility of alternative treatment of alopecia (Young et al. 2014).

2.3.9 Emblica Officinalis

The dried as well as fresh fruits of *Emblica officinalis*, belonging to family Euphorbiaceae are used. It is found in the deciduous forests in India and distributed in the tropical and subtropical zones. It is one of the important ingredient of *Ayurvedic* preparations: 'Trifala' and 'Chyawanprash'. The oil is extracted by solvent extraction using petroleum ether.

Chemical constituents
It is the richest source of Vitamin C. It contain minerals like phosphorous, calcium and iron, phyllemblin, tannins and amino acids like glutamic acid, aspartic acid, proline, alanine and lysine. The oil contains linoleic, linolenic, oleic acids and myristic, palmitic and stearic acids (Dhar et al. 1951).

Benefits to hair
The fixed oil obtained from the fruits possess hair growth promoting activity. The dried fruits are boiled in coconut oil, application of this oil prevents graying of hair (Kumar et al. 2012). It can also be used to cleanse the hair, to control dandruff, to prevent hair fall (Sharma and Agarwal 2003).

2.3.10 Hibiscus rosa sinensis

Hibiscus rosa sinesnis is a flowering plant in the mallow family, Malvaceae, indigenous to warm-temperate, subtropical and tropical regions throughout the world.

Chemical constituents
Leaves and stems contain β-sitosterol, stigma sterol, taraxeryl acetate and three cyclo propane compounds and their derivatives. Flowers contain cyanidine diglucoside, flavonoids and vitamins, thiamine, riboflavin, niacin and ascorbic acid (Rao et al. 2014).

Benefits to hair
Hibiscus flower acts as a wonderful remedy for hair loss problem since this flower has a great combination of vitamin C, phosphorous, riboflavin and calcium. It also helps to soften the hair. It prevents pre mature graying of hair due to the carotene content. It has a very mild foaming effect and hence can be incorporated in baby shampoos. The leaves and flowers on burning give a black dye which can be used to color hair.

2.3.11 Juniperus communis (Juniper Berry)

It is a tree belonging to the family Cupressaceae, with a wide distribution ranging from the Arctic region, Africa, Central America to Pakistan. It is a cone with fleshy and merged scales giving a berry like appearance.

Chemical composition
They contain volatile oil as the main constituent, invert sugar and resin. These are rich source of antioxidants and are proved to be anti-infective. The volatile oil comprises of monoterpene hydrocarbons such as α-pinene, β-myrcene, sabinene, limonene, terpine-4-ol, β-caryophyllene and β-pinene.

Benefits for the hair
Massaging the scalp and hair at least 2 h prior to washing with a blend of juniper,
rosemary and cedarwood essential oils in olive oil, helps to stimulate circulation to
the scalp, promotes scalp health and also provides moisturizing effect.

2.3.12   Oenothera biennis *(Evening Primrose)*

This is a species of Oenothera native to eastern and central North America.
Oenothera biennis is a biennial plant, producing flowers on a tall spike. Since they
open in the evening the plant is given the name evening primrose.

Chemical constituents
It contains fatty acids, including non-essential ω-6 polyunsaturated fatty acid,
γ-linolenic acid, oleic acid, palmitic acid and stearic acid. It also contains steroids:
campesterol and β-sitosterol and 3-*O*-trans-caffeoyl esters of triterpene acids
(betulinic, morolic, and oleanolic acids).

Benefits to hair
Gamma-linolenic acid is an omega-6 fatty acid, which is one of the essential fatty
acids necessary for healthy development and human growth. The same component
provides nourishment to the scalp and hair and may be incorporated in hair oils or
may be used as a carrier oil.

2.3.13   Rosmarinus officinalis *(Rosemary)*

The plant is commonly known as rosemary, belonging to the family Lamiaceae
(Genena et al. 2008). It is a spice and medicinal herb widely accepted as one of the
spices with the highest antioxidant activity (Jiang et al. 2011).

Chemical constituents
Rosemary is reported to have flavones, diterpenes, steroids and triterpenes. The
antioxidant activity of rosemary is because of carnosic acid and carnosol. α-pinene,
bornyl acetate, camphor and 1,8-cineole are the main compounds responsible for
the antimicrobial activity (Jiang et al. 2011; Murata et al. 2013).

Benefits to hair
Topical application of leaf extract reverses the interruption in hair growth induced
by testosterone treatment in the mice. Extract of *Rosmarinus officinalis* show
antiandrogenic activity. 12-methoxycarnosic acid, present in rosemary, inhibited
5α-reductase enzyme and has been identified as the active constituent. Thus it may
be used in hair oils to prevent hair loss (Murata et al. 2013).

2.3.14   Solanum tuberosum

It is commonly known as potato and is a rich source of starch from the Solanaceae family. The tubers are used as a part of diet and also for beneficial effects on hair.

Chemical constituents
Potato is rich in Vitamin A, B, C and D. These are essential for healthy hair and can be useful even if one is suffering from alopecia. They also contain protein, starch, iron, fiber, potassium, magnesium, phosphorous, calcium and possess hydrophillic antioxidants (Mouillé et al. 2010; Brautlecht and Getchell 1951).

Benefits to hair
Being a rich source of vitamins, if potato juice is incorporated in the oils, it will not only prevent hair loss but may also prevent the hair from becoming dry and brittle.

2.3.15   Trigonella foenum-graecum *(Fenugreek)*

It is an annual plant in the family Fabaceae, with leaves consisting of three small obovate to oblong leaflets. It is cultivated worldwide as a semiarid crop, and its seeds are a common ingredient in dishes from the Indian Subcontinent. It was well known since ancient times and the seeds were used medicinally.

Chemical constituents
Diosgenin, is the main steroid found in fenugreek. Other sapogenins found in fenugreek seed include yamogenin, gitogenin, tigogenin, and neotigogens. Fenugreek seeds contain alkaloids, including trigonelline, gentianine and carpaine compounds. The seeds also contain fiber, 4-hydroxyisoleucine and fenugreekine, a component that may have hypoglycemic activity. Other constituents of fenugreek include mucilage, proteins, bitter fixed oil, volatile oil, and the alkaloids choline and trigonelline. It a rich source of protein, Biotin, iron, Vitamin A, Vitamin B1, Vitamin B2, Vitamin B3, Vitamin B5, Vitamin B6, Vitamin B9, Vitamin B12 and Vitamin D (Mullaicharam et al. 2013).

Benefits to hair
Fenugreek is a potent dihydrotestosterone (DHT) blocker and rich source of multivitamin B which plays major role in hair stimulant property. DHT is synthesized in adrenal glands, hair follicles, testes and prostate by the action of the enzyme 5α-reductase, which plays a major role in occurrence of alopecia. Both males and females are affected, but men commonly being affected more. Blocking this DHT synthesis helps in preventing hair loss and treating alopecia. Thus fenugreek seed extract could be incorporated in oils or shampoos for hair loss prevention.

References

Ahmad A, Husain A, Mujeeb M et al (2013) A review on therapeutic potential of *Nigella sativa*: A miracle herb. Asian Pac J Trop Biomed 3(5):337–352

Alpaslan M, Hayta M (2006) Apricot kernel: physical and chemical properties. J Am Oil Chem Soc 83(5):469–471

Amagase H (2006) Clarifying the real bioactive constituents of garlic. J Nutr 136(3):716S–725S

Angadi R (2009) A text book of Bhaisajya kalpana vinjana. Chaukambha Surbharati Prakshan, Varanasi

Anwar F, Latif S, Ashraf M (2006) Analytical characterization of hemp (*Cannabis sativa*) seed oil from different agro-ecological zones of Pakistan. J Am Oil Chem Soc 83(4):323–329

Appaiah P, Sunil L, Prasanth Kumar PK (2014) Composition of coconut testa, coconut kernel and its oil. J Am Oil Chem Soc 91(6):917–924

Barnes PJ (1982) Lipid composition of wheat germ and wheat germ oil. Fette Seifen Anstrichm 84:256–269. doi:10.1002/lipi.19820840703

Bosy TZ, Cole KA (2000) Consumption and quantitation of Δ9-tetrahydrocannabinol in commercially available hemp seed oil products. J Anal Toxicol 24(7):562–566

Brautlecht CA, Getchell AS (1951) The chemical composition of white potatoes. Am Potato J 28(3):531–550

Burlando B, Verotta L, Cornara L et al (2010) Herbal principles in cosmetics: properties and mechanisms of action. CRC press, Taylor and Francis group, Boca Raton

Callaway JC (2004) Hempseed as a nutritional resource: an overview. Euphytica 140(1):65–72

Choi JS, Jeon MH, Moon WS (2014) In vivo hair growth-promoting effect of rice bran extract prepared by supercritical carbon dioxide fluid. Biol Pharm Bull 37(1):44–53

Chowdhury K, Banu LA, Khan S et al (2007) Bangladesh. J Sci Ind Res 42(3):311–316

Cicero AF, Gaddi A (2001) Rice bran oil and y—oryzanol in the treatment of hyperlipoproteinaemias and other conditions. Phytotherapy Res 15(4):277–289

Dhar DC, Dhar ML, Shrivastava DL (1951) Chemical examination of the seeds of *Emblica officinalis* Gaertn.: Part I—The fatty oil & its component fatty acids. Sci Industr Res 10B: 88–91

Dini I, Tenore GC, Dini A (2008) Chemical composition, nutritional value and antioxidant properties of *Allium caepa* L. Var. Tropeana (red onion) seeds. Food Chem 107:613–621

Dwivedi DH, Ram RB (2008) Chemical composition of bitter apricot kernels from Ladakh, India. Acta Hortic 765:335–338

Ekin Z (2005) Resurgence of safflower (*Carthamus tinctorius* L.) utilization: a global view. J Agron 4(2):83–87

Ferrari RA, Soler MP (2015) Obtention and characterization of coconut babassu derivatives. Sci Agric (Piracicaba, Braz.), Piracicaba 72(4):291–296. Available from http://www.scielo.br/scielo.php?script=sci_arttext&pid=S0103-90162015000400291&lng=en&nrm=iso. Accessed on 30 Nov 2015. http://dx.doi.org/10.1590/0103-9016-2014-0278

Fife B (2013) The coconut oil miracle. Penguin group, New York

Genena AK, Haiko H, Junior AS et al (2008) Rosemary (*Rosmarinus Officinalis*)—a study of the composition, antioxidant and antimicrobial activities of extracts obtained with supercritical carbon dioxide. Ciênc Tecnol Aliment Campinas 28(2):463–469

Graham HN (1992) Green tea composition, consumption, and polyphenol chemistry. Prev Med 21(3):334–350

Guillaume D, Charrouf Z (2011) Argan oil. Altern Med Rev 16(3):275–279

Gupta A, Sharma PC (2009) Standardization of technology for extraction of wild apricot kernel oil at semi-pilot scale. Biol Forum Int J 1(1):51–64

Jiang Y, Wu N, Fu YJ et al (2011) Chemical composition and antimicrobial activity of the essential oil of rosemary. Environ Toxicol Pharmacol 32(1):63–68

Keis K, Huemmer CL, Kamath YK (2007) Effect of oil films on moisture vapor absorption on human hair. J Cosmet Sci 58(2):135–145

Khallouki F, Younos C, Soulimani R et al (2003) Consumption of argan oil (Morocco) with its unique profile of fatty acids, tocopherols, squalene, sterols and phenolic compounds should confer valuable cancer chemopreventive effects. Eur J Cancer Prev 12(1):67–75

Knowles J, Watkinson C, Johnstone I (2014) Extraction of wheatgerm: the production of wheatgerm oil and de-fatted stabilised wheatgerm. Lipid Technol 26:157–161

Kumar P, Yadava RK, Gollen B (2011) Nutritional contents and medicinal properties of wheat: a review. Life Sci Med Res LSMR-22

Kumar S, Swarankar V, Sharma S et al (2012) Herbal cosmetics: used for skin and hair. Inventi Rapid Cosmeceuticals 2012(4):1–7

Lee YC, Oh SW, Chang J et al (2004) Chemical composition and oxidative stability of safflower oil prepared from safflower seed roasted with different temperatures. Food Chem 84(1):1–6

Li Z et al (2006) Extraction in almond oil. J Anhui Agric Sci

Marina AM, Che Man YB, Nazimah SAH et al (2009) Chemical properties of virgin coconut oil. J Am Oil Chem Soc 86(4):301–307

McGlone OC, Canales AL, Carter JV (1986) Coconut oil extraction by a new enzymatic process. J Food Sci 51(3):695–697

Megahad OA, El Kinawy OS (2002) Studies on the extraction of wheat germ oil by commercial hexane. Grasas Aceites 53(4):414–418

Miyeon K, Kyoungsook K (2013) The effect of rosehip extracts addition on permanent wave and hair dye during repetition procedure. Fashion Bus 17(2):151–163

Monfalouti HE, Guillaume D, Denhez C et al (2010) Therapeutic potential of argan oil: a review. J Pharm Pharmacol 62(12):1669–1675

Mouillé B, Charrondière UR, Burlingame B (2010) Nutrient composition of the potato interesting varieties from human nutrition perspective. Food And Agriculture Organization of the United Nations, Rome, Italy, http://www.fao.org/fileadmin/templates/food_composition/documents/upload/Poster_potato_nutrient_comp.pdf. Accessed on 27 October 2015

Mullaicharam AR, Deori G, Maheswari RU (2013) Medicinal values of fenugreek—a review. Res J Pharm Biol Chem Sci 4(1):1304–1313

Murata K, Noguchi K, Kondo M et al (2013) Promotion of hair growth by *Rosmarinus officinalis* leaf extract. Phytother Res 27(2):212–217

Nagendra Prasad MN, Sanjay KR, Khatokar SM (2011) Health benefits of rice bran—a review. J Nutr Food Sci 1:3. Accessed on 30th Nov 2015. http://dx.doi.org/10.4172/2155-9600.1000108

Patel M, Naik SN (2004) Gamma-oryzanol from rice bran oil—a review. J Sci Ind Res 63:569-578

Procida G, Stancher B, Cateni F et al (2013) Chemical composition and functional characterisation of commercial pumpkin seed oil. J Sci Food Agric 93(5):1035–1041

Putt ED, Carson RB (1969) Variation in composition of sunflower oil from composite samples and single seeds of varieties and inbred lines. J Am Oil Chem Soc 46(3):126–129

Rao KNV, Geetha K, Raja A et al (2014) Quality control study and standardization of *Hibiscus rosasinensis* I. flowers and leaves as per WHO guidelines. J Pharm Phytochem 3(4):29–37

Rele A, Mohile R (2003) Effect of mineral oil, sunflower oil, and coconut oil on prevention of hair damage. J Cosmet Sci 54(2):175–192

Ruan CJ, Teixeira da Silva JA, Jin H et al (2007) Research and biotechnology in sea Buckthorn (*Hippophae* spp.). Med Aromat Plant Sci Biotechnol 1(1):47–60

Rueda A, Seiquer I, Olallca M et al (2014) Characterization of fatty acid profile of Argan oil and other edible vegetable oils by gas chromatography and discriminant analysis. J Chem. http://dx.doi.org/10.1155/2014/843908

Sagarin E (2008) Cosmetics, science and technology. Wiley Interscience, Wiley India Pvt. Ltd., New Delhi

Servili M, Esposto S, Fabiani R eta al (2009) Phenolic compounds in olive oil: antioxidant, health and organoleptic activities according to their chemical structure. Inflammopharmacology 17(2):76–84

Sharma L, Agarwal GA (2003) Medicinal plants for skin and hair care. IJTK 02(1):62–68

Stevenson DG, Eller FJ, Wang L et al (2007) Oil and tocopherol content and composition of pumpkin seed oil in 12 cultivars. J Agric Food Chem 55(10):4005–4013

Suryakumar G, Gupta A (2011) Medicinal and therapeutic potential of Sea buckthorn (*Hippophae rhamnoides* L.). J Ethnopharmacol 138(2):268–278

Szentmihályi K, Vinkler P, Lakatos B (2002) Rose hip (*Rosa canina* L.) oil obtained from waste hip seeds by different extraction methods. Bioresour Technol 82(2):195–201

Tripathi V, Abidi AB, Marker S et al (2013) Linseed and linseed oil: health benefits—a review. Int J Pharm Biol Sci 3(3):434–442

Unna K, Sampson WL (1940) Effect of pantothenic acid on the nutritional achromotrichia. Proc Soc Exp Biol Med 45:309–311

Vaidya B, Cho E (2011) Effects of seed roasting on tocopherols, carotenoids, and oxidation in mustard seed oil during heating. J Am Oil Chem Soc 88(1):83–90

Waller GR, Mangiafico S, Ritchey CR (1978) A chemical investigation of *Aloe barbadensis* Miller. Okla Acad Sci 58:69–76

Wallis TE (2005) Textbook of pharmacognosy. CBS Publishers and Distributors Pvt. Ltd., Delhi

Wisniak J (1987) The chemistry and technology of Jojoba oil American oil chemists society press. Champagn, Illinois, USA

Yong JWH, Ge L, Fei NY et al (2009) The chemical composition and biological properties of coconut (*Cocos nucifera* L.). Water Mol 14:5144–5164

Young HC, Sang YL, Dong WJ et al (2014) Effect of pumpkin seed oil on hair growth in men with androgenetic alopecia: a randomized, double-blind, placebo-controlled trial. Evidence-based complementary and alternative medicine 2014. http://dx.doi.org/10.1155/2014/549721

Zhong J, Huang Y, Ding W et al (1991) Chemical constituents of *Aloe barbadensis* Miller and their inhibitory effects on phosphodiesterase-4D. Fitoterapia:159–165

Chapter 3
Hair Shampoo

Abstract Regular use of soap or foam forming agents gives a satisfaction of cleaning one's body. The same is applicable for the scalp to more or less extent, the only difference being in the frequency of using such agents for cleaning. Cleansing agents for scalp and hair are termed as shampoos. Shampoos are used at least two-three times a week for washing hair. The foam-ability is imparted by surfactants. These surfactants have lot of side effects and hence the chapter provides information on plant sources which could be used as alternative surfactants. Few plant drugs having benefits of removing lice or dandruff, healing of scalp wounds are described which may be incorporated in shampoos.

Keywords Saponins · Surfactants · Shikakai · Reetha · Tea tree oil · Amla · Yucca

3.1 Introduction

One of the largest commercial units in the cosmetic industry is the hair care sector. Shampoos are used primarily to cleanse the hair and the scalp from the accumulated sebum, scales, scalp debris and residues of hair-grooming preparations. It also helps in auxiliary functions like providing medication, lubrication, body building and prevention of static charge build (Arora et al. 2011).

When we need to formulate a shampoo, one has to consider the ideal characteristics along with the ingredients to be incorporated in the shampoo.

Ideal characteristics of shampoo (Preethi and Padmini 2013):

1. It should completely remove dirt
2. Good foaming property
3. It should be easily removed after rinsing
4. It should make hair soft lustrous and manageable
5. It should not make hands dry and chapped
6. It should not cause irritation to eyes and skin.

© The Author(s) 2016
K. Barve and A. Dighe, *The Chemistry and Applications of Sustainable Natural Hair Products*, SpringerBriefs in Green Chemistry for Sustainability, DOI 10.1007/978-3-319-29419-3_3

Composition of shampoo (Preethi and Padmini 2013):

1. Principle surfactant
2. Secondary surfactant
3. Antidandruff agent (if required)
4. Conditioning agent
5. Pearlescent agent
6. Thickening agent
7. Colours, perfume and preservative

Surfactants dissolve the impurities and residues on hair and prevent them from binding to the shaft or the scalp. The surfactants are classified in four groups depending on the charge: Anionic, cationic, amphoteric and nonionic. Anionic surfactants act as cleansing agents. The other classes of surfactants are added to reduce the static electricity generating effects caused by the anionic surfactants and optimize the formation of foam and viscosity of the final product (Dias 2015).

Sodium lauryl sulphate (SLS) is the most commonly used surfactant in shampoos but it is rather a harsh detergent. Long term use of SLS may cause problems in the heart, liver, lungs and brain. It may impair the ability of hair to grow due to corrosion of hair follicle. Cocamidopropyl betaine is another commonly used surfactant in shampoos. It is mostly used in no tears shampoos for children and there are many reported cases for dermatitis. Diethanolamine and N-nitrosodiethanolamine (NDELA) are also present in shampoos and may pose a serious health risk to consumers.

Looking into the side effects of synthetic ingredients used in shampoos, we need to use natural materials, being produced from plants, which are both bio-degradable and non-toxic (Pradhan and Bhattacharyya 2014).

Shampoos prepared only with natural products may not be harmful to that extent, either to the user or to the environment (Pradhan and Bhattacharyya 2014). Natural constituents like saponins which act as foaming agent can be used as mild cleaning agents too. Saponins are found in abundance in plants like a Shikakai (*Acacia concinna*), Fenugreek (*Trigonella foenum-graecum*) etc.

Shampoos based on herbal ingredients

Active principles of various plants like phyto-hormones, flavonoids, hydroxyl acids also referred to as fruit acids, glycosides-saponins, enzymes, and essential oils are being considered useful in cosmetic formulations. Saponins act as a main ingredient as it will help to generate foam and clean the scalp.

Saponins

Deriving their name from the Latin sapo, meaning soap, saponins have long been implicated as that plant constituent producing frothing in aqueous solution also referred to as a "natural detergent". Chemically, the term saponin has become accepted to define a group of structurally diverse molecules that consists of glycosylated steroids, steroidal alkaloids and triterpenoids. Properties shared by this group of natural products are (Ullmann 1996):

1. Surfactant activity: Combination of hydrophobic aglycone and hydrophilic substituents accounts for the amphiphilic nature of saponins. They lower the surface tension of aqueous solutions and therefore give stable foams when in contact with water.
2. Hemolytic action: Depends on type and structure of substituents. Bisdesmosides are generally less potent than monodesmosides.
3. Steroid complexing ability: More with steroid saponins and glycoalkaloids than with triterpenoids.
4. Biocidal capability: Permeabilization of crucial cell membrane probably explains their cell toxicity.

Saponin containing plants are sought after use in household detergents (Sparg et al. 2004). One such example is the soapwort (*Saponaria officinalis*), which was widely used for centuries. Triterpenoid saponins have been used in the manufacture of foam fire extinguishers, toothpaste, and foam in beverages (including soft drinks and beer), shampoos, liquid soaps, and cosmetic preparations.

Below is mentioned a list of plants which act as cleansing agent and may be incorporated in the shampoo.

3.2 *Acacia concinna* (Shikakai)

It is found in Asia and is a rich source of saponins. Since ancient times it has been used traditionally for hair care. Extract of this plant is used in the herbal shampoos and hair powder. The fruit pods, barks and leaves are dried, ground and made into a paste and then applied to hair as a shampoo. It does not alter the hair texture since it has a low pH and is also mild in its cleansing effect. Bark contains high levels of saponins which make them lather producing.

Chemical constituents
The acacia pods contain several saponins. Recently acacic acid lactone 3β-acetate and a new nor-triterpene 'acacidiol' have been isolated from the pods (Anjaneyulu et al. 1979).

Benefits for the hair
Shikakai is very mild in nature and is used for cleaning and conditioning the hair. It has to be noted that a natural shampoo made from Shikakai does not make much lather like the other shampoos but it is a much better and effective way of cleaning hair. There is considerably low pH found in Shikakai which means that when used as a shampoo it will not leave the hair without its natural oils. Shikakai is also an excellent natural conditioner and detangles the hair without any side effects. It also possesses astringent characteristics and they enable the herb to remove all the dirt and unwanted oil collected on the scalp. It is an effective remedy for hair lice as well as dandruff. Shikakai is used to clean hair after dyeing as it helps to absorb hair

dye in a much better manner. These qualities of Shikakai help in keeping the hair healthy and lustrous and it also helps in preventing hair fall thus enabling the hair to grow in length and volume. When used in the summer season, it helps the body to stay fresh and cool and also retains the needed moisture on scalp.

3.3   *Artemissia abrotanum* (Southernwood)

Artemissia abrotanum is a species of flowering plants in the sunflower family. It is native to Eurasia and Africa but naturalized in scattered locations in North America. Southernwood has a strong camphor-like odour and was historically used as an air freshener or strewing herb.

Chemical composition
It contains phenolic acids such as caffeic, ferulic, vanillic and salicylic acids, abrotanin, flavonols and volatile oils. The volatile oil mainly comprise of methyl isopropyl ketone, eucalyptol, acetophenone, linalool, alpha copane, eugenol, iso eugenol and cis-jasmone (Tunón et al. 2006).

Benefits for the hair
It can be combined with Rosemary as a hair rinse to encourage the growth of hair and improve hair condition. The volatile oil may also be beneficial in removing hair lice since it has already been reported as a repellant for mosquitoes and ticks.

3.4   *Azadirachta indica* (Neem)

It is a tree belonging to the family Meliaceae also known as the Mahogany family. It is also known as Neem or Indian Lilac. It is one of two species in the genus Azadirachta, and is native to India.

Chemical composition
Active constituents of neem leaf extract include isomeldenin, nimbin, nimbinene, 6-desacetyllnimbinene, nimbandiol, nimocinol, quercetin, and beta-sitosterol. Two additional tetracyclic triterpenoids zafaral [24,25,26,27-tetranorapotirucalla-(apoeupha)-6alpha-methoxy-7alpha-acetoxy-1,14-dien-3,16-dione-21-al] and meli-acinanhydride   [24,25,26,27-tetranorapotirucalla-(apoeupha)-6alpha-hydroxy,11 alpha-methoxy-7alpha,12alpha-diacetoxy,1,14,20(22)-trien-3-one] have been iso-lated from the methanolic extract of neem leaves. It can be incorporated as an ingredient in the shampoo due to the below mentioned benefits.

Benefits for the hair
Neem leaf is used to treat the problem of head lice, dandruff, and excessive hair fall and to improve hair growth. Application of paste of boiled neem leaves on the scalp,

prevents the problem of head lice. This treatment controls hair fall, prevents premature greying and also boosts hair growth.

3.5   *Chlorogalum pomeridianum* (Amoly Lilly)

The common names Soap Plant, Soap root or Amole refer to the genus Chlorogalum. They are native to western North America, from Oregon to Baja California, and are mostly found in California. Soap Plants are perennial plants, with more or less elongated bulbs, depending on the species. The bulbs can be white or brown, and in most species have a fibrous coat.

Chemical composition
They are rich source of saponins.

Benefits for the hair
Since this is rich source of saponins which are foaming agents it can be used as hair cleanser.

3.6   *Citrus Limon* (Lemon)

Citrus Limon is a species of small evergreen tree native to Asia. The juice from fruits may be used for cleaning hair. The low pH of juice makes it antibacterial. Lemon oil derived from the rind of the fruits is used in aromatherapy to enhance mood.

Chemical composition
The volatile oil mainly contains limonene (about 90 %), citral and other aromatic compounds like citronellal, geranyl acetate, linalyl acetate, octyl and nonyl aldehydes and terpineol.

Benefits for the hair
Being a source of vitamin B and C, it provides nourishment to the hair follicles and promotes their growth and prevents hair fall. As mentioned above due to its antibacterial effect lemon juice also helps in controlling dandruff and prevents its recurrence. Application of crushed lemon seeds and black pepper as a paste or equal parts of lemon and vinegar clear the clogged follicles by reducing the excessive secretion of sebum, paving way for nourishing the follicle and the shaft and preventing hair fall. It has natural bleaching properties and may be used for lightening the hair. Garlic or almond paste may be added to the juice, which could be an effective treatment for the lice. Application of fresh coconut water with lemon juice to the hair scalp, gives a shine and luster to hair.

3.7   *Hemidesmus indicus*/Indian Sarsaparilla (Sariva, Sugandhi, Uparsal, Ananthamoola)

It is a species of plant that is found in South Asia. It is a slender, laticiferous, twining, sometimes prostrate or semi-erect shrub. Roots are woody and aromatic. The stem is numerous, slender, terete, thickened at the nodes. The leaves are opposite, short-petioled, very variable, elliptic-oblong to linear-lanceolate. The flowers are greenish outside, purplish inside, crowded in sub-sessile axillary cymes. It is occurs over the greater part of India, from the upper Gangetic plain eastwards to Assam and in some places in central, western and South India (George et al. 2008).

Chemical constituents (Moideen et al. 2011; Rajan et al. 2011)
Coumarins, flavonoids, tannins, alkaloids, saponins and steroids are present in the leaves. The roots also show a similar profile of phytoconstituents. The flavanoid glycosides recognized in the flowers are hyperoside, isoquercitin and rutin whereas in the leaves, only hyperoside and rutin are identified. Some of the many compounds found in this plant include: 2-hydroxy-4-methoxy benzaldehyde, 2-hyroxy-4-methoxy benzenoid, alpha-amyrins triterpene, benzoic acid, beta-amyrins. The major volatile components present are 2-hydroxy,4-methoxy benzaldehyde and ledol. Two glycosides namely indicine and hemidine have been isolated from the stem (George et al. 2008).

Benefits for hair
The herb is incorporated in hair oils as a hair tonic to promote hair growth (Punjani and Kumar 2003). The local people apply the paste of the root on the scalp to promote hair growth (Pawar and Patil 2012).

3.8   *Melaleuca alternifolia* (Tea Tree)

Tea tree oil or Melaleuca oil is an essential oil with a fresh camphoraceous odour and a color that ranges from pale yellow to nearly colorless and clear. It is taken from the leaves of the *Melaleuca alternifolia*, which is native to Southeast Queensland and the Northeast coast of New South Wales, Australia.

Chemical composition
Tea tree oil contains over 98 compounds, and has six chemo types, which are oils with different chemical compositions. These include a terpinen-4-ol chemo type, a terpinolene chemo type, and four 1, 8-cineole chemo types. Terpinen-4-ol is the major tea tree oil component responsible for antimicrobial and anti-inflammatory properties. This can be useful for scalp having some kind of infections.

Benefits for the hair
Tea tree oil has the ability to deplete ATP (adenosine trisphosphate) levels and thus prevent ATP dependant pesticide resistance, and hence is incorporated in shampoos meant for the removal of head lice (McCage et al. 2002). It also shows effectiveness in the treatment of dandruff and thus can be an ingredient in one of anti-dandruff shampoos (Satchell et al. 2002).

3.9 *Phyllanthus emblica* (Amla)

The plant is also known as emblica, Indian gooseberry or Amla. The tree is small to medium in size with a crooked trunk and spreading branches. The branchlets are glabrous or finely pubescent, usually deciduous; the leaves are simple, subsessile and closely set along branchlets, light green, resembling pinnate leaves. The berries which ripen in autumn are harvested by hand after climbing to upper branches bearing the fruits. The taste of Indian gooseberry is sour, bitter and astringent, and it is quite fibrous.

Chemical compositions
Amla fruit is rich in Vitamin C (ascorbic acid), is a source of invaluable minerals such as calcium, magnesium, potassium, iron, copper and more, as well as a source of some amino acids (alanine, arginine, aspartic acid and others). Additionally, it is abundant in polyphenols, particularly tannins that are derived from gallic acid and ellagic acid, including emblicanin A and B, punigluconin and pedunculagin. It also contains flavonoid such as rutin. The seeds yield a fixed oil which is brownish-yellow in colour. It has the following fatty acids: linolenic, linoleic, oleic, stearic, palmitic and myristic acid.

Benefits for the hair
Amla-Berry boosts absorption of calcium, thus creating healthier bones, teeth, nails and hair. It also helps maintain youthful hair color and retards premature graying, and supports the strength of the hair follicles, so there is less thinning with age. The crushed fruits have a good effect on hair growth and prevent hair graying (Singh et al. 2011).

3.10 *Sapindus mukorossi* (Reetha)

It is commonly known as Reetha and belongs to the Sapindaceae family. The fruit is also referred to as a washnut or soapberry. The trend of washing hair with soapnut (reetha) is still followed in many local households. The fruits contain saponin which are natural surfactants and can be used as a cleanser for hair, skin, and clothing. These saponins are also useful as insecticides, for purposes such as removing head lice from the scalp.

Chemical composition
The pericarp of the fruit contains 10–11 % of saponins. Major constituents present in the fruits are saponins, triterpenoids, fatty acids and flavonoids. Some other species have astringent characteristics ingredients. These ingredients are well known for antimicrobial activity, fungicidal and inti-inflammatory which will be useful for hair products. Solubilising property was observed in Mukurozi saponins-Monodesmoides and bisdesmoides isolated from pericarps (Sharma et al. 2011).

Benefits for the hair
Reetha is used as the main ingredient in soaps and shampoos for washing hair, as it is considered good for the health of hair. It has been placed as a popular herb in the list of herbs and minerals in *Ayurveda* and is used as an important ingredient in cleansers and shampoos. In addition, it is used for the treatment of eczema, psoriasis, and for removing freckles. It is also used for removing lice from the scalp, as it has gentle insecticidal properties. The plant is known for its antimicrobial properties that are beneficial for wounds on scalp.

3.11   *Saponaria officinalis* (Soapwort)

Saponaria officinalis is a common perennial plant from the carnation family, Caryophyllaceae. This plant has many common names, including common soapwort, bouncing-bet, crow soap, wild sweet William, and soap weed. There are about 20 species of soapworts altogether (Zhonghua et al. 1998).

Chemical compositions
Two major triterpenoid saponins, named saponariosides A and B, six novel triterpenoid saponins, named saponariosides C–H, were isolated from the whole plants of *Saponaria officinalis* (Zhonghua et al. 1999).

Benefits for the hair
Soapwort is rich source of saponins which produce much foam with water. It helps to clean hair without damaging its structure. It can be combined with horsetail (*Equisetum arvense*), contains silicon that will help to protect hair from falling.

3.12   *Quillaja saponaria* (Soap Bark)

Quillaja saponaria, the soap bark tree or Soapbark, is an evergreen tree in the family Quillajaceae, native to warm temperate central Chile. The inner bark of *Quillaja saponaria* can be reduced to powder and employed as a substitute for soap, since it forms lather with water, owing to the presence of a glucoside saponin, sometimes distinguished as quillaia saponin.

Chemical composition
Quillaja extract contain over 100 triterpenoid saponins, consisting predominantly of glycosides of quillaic acid. Polyphenols and tannins are also major components. Quillaja triterpenoidal saponins are non-ionic surfactants, resistant to salt, heat, and extremely stable to acid pH.

Benefits for the hair
Using *Quillaja saponaria* externally is believed to soothe itchy skin and scalp irritation. It is also used to treat dandruff and skin ulcers as it is a gentle cleanser.

3.13 *Yucca angustifolia* (Yucca)

Yucca is a genus of perennial shrubs and trees in the family Asparagaceae, sub-family Agavoideae, also colloquially known as "ghosts in the graveyard", as it is commonly found growing in rural graveyards. Yucca plants have served indigenous people for centuries for a variety of uses, including fiber for rope, sandals, and cloth; the roots have been used in soap. More recently, yucca has been used in soaps, shampoos, and food supplements. Yuccas contain saponins that have a long-lasting foaming action.

Chemical composition
The roots and flowers of the yucca are rich in steroidal saponins consisting of sarsasapogenin and tigogenin as the aglycone. Phenolic compounds (novel yuccaols and gloriosaols) are also numerous in yucca species.

Benefits for the hair
It is a Native American remedy that may be used for hair loss. The roots have been proven extremely effective and may be included in several shampoos recommended for people with hair thinning or hair loss. The extract contains saponins, which is a cleanser that will also reduce inflammation and itchiness. The saponins may also be effective in reducing dandruff, which may be considered as a cause of hair loss in some people. Yucca is not a cure for baldness or hair loss, but the root extracts can improve the quality of the remaining hair and prevent further hair loss.

3.14 Natural Surfactants

There are few excellent surfactants which are derived and isolated from fruits or vegetable sources can be used in shampoos.

Apple Surfactant
It is an anionic biodegradable surfactant derived from apple juice and being mild in nature is used in hair products for sensitive hair and baby shampoos.

Cocamidopropyl Betaine
It is a mild, amphoteric natural surfactant derived from coconut oil, which gives good foam characteristic to the formulation and improves viscosity.

DLS Mild
DLS Mild is a natural surfactant derived from natural Coconut or Palm Kernel oil and is used in baby shampoos and helps to maintain the natural oils in hair.

Plantapon
Plantapon® LGC SORB is an anionic surfactant derived from a reaction between coconut/palm oil and starch/sugar. It improves the foaming properties and makes the hair soft.

PlantaSol CCG
PlantaSol CCG is a glucolipidic (sugar-based) surfactant helps in building viscosity and boosts foaming. It can also easily dissolve essential oils in shampoos.

SCI 85 % Noodle
Sodium Cocoyl Isothioniate (SCI), based on fatty acids from coconut oil, offers dense, luxurious foam in hard or soft water and is biodegradable.

SCI Pearl Surfactant
Sodium cocoyl isethionate is an anionic surfactant derived from coconut, gives food foam and is used to make mild shampoo.

References

Anjaneyulu ASR, Row LR, Sree A (1979) Acacidiol, a new nor-triterpene from the sapogenins of *Acacia concinna*. Phytochemistry 18(7):1199–1201

Arora P, Nanda A, Karan M (2011) Shampoos based on synthetic ingredients vis-a-vis shampoos based on herbal ingredients: a review. Int J Pharm Sci Rev Res 7(1):42

Dias M (2015) Hair cosmetics: an overview. Int J Trichology 7(1):2–15

George S, Tushar KV, Urmikrishnan KP (2008) *Hemidesmus indicus* (L.) R. Br. A review. J Plant Sci 3(2):146–156

McCage CM, Ward SM, Paling CA et al (2002) Development of a paw paw herbal shampoo for the removal of head lice. Phytomedicine 9(8):743–748

Moideen MM, Varghese R, Krishna Kumar E (2011) Wound healing activity of ethanolic extract of *Hemidesmus indicus* (Linn) R. Br leaves in rats. Res J Pharm Biol Chem Sci 2(3):643–651

Pawar S, Patil DA (2012) Herbal haircare as revealed by people in Jalgaon district, Maharashtra, India. J Exp Sci 3(3):32–34

Pradhan A, Bhattacharyya A (2014) Shampoos then and now: synthetic versus natural. J Surf Sci Technol 30(1–2):59–76

Preethi J, Padmini K (2013) A review on herbal shampoo and its evaluation. Asian Pharma Press 3:153–156

Punjani BL, Kumar V (2003) Plants used in traditional phytotherapy for hair care by tribals in Sabarkantha district, Gujarat, India. IJTK 02(1):74–78

Rajan S, Shalini R, Bharathi C et al (2011) Pharmacognostical and phytochemical studies on *Hemidesmus indicus* root. Int J Pharmacognosy Phytochem Res 3(3):74–79

Satchell AC, Saurajen A, Bell C et al (2002) Treatment of dandruff with 5 % tea tree oil shampoo. J Am Acad Dermatol 47(6):852–855

Sharma A, Sati SC, Sati OP et al (2011) Chemical constituents and bioactivities of genus sapindus. Int J Res Ayurveda Pharm 2(2):403–409

Singh E, Sharma S, Pareek A et al (2011) Phytochemistry, traditional uses and cancer chemopreventive activity of Amla (*Phyllanthus emblica*): the sustainer. J Appl Pharma Sci 2(1):176–183

Sparg SG, Light ME, Staden JV (2004) Biological activities and distribution of plant saponins. J Ethnopharmacol 94(2–3):219–243

Tunón H, Thorsell W, Mikiver A et al (2006) Arthropod repellency, especially tick (*Ixodes ricinus*), exerted by extract from *Artemisia abrotanum* and essential oil from flowers of *Dianthus caryophyllum*. Fitoterapia 77(4):257–261

Ullman F (1996) Ullmans encyclopedia of industrial chemistry. A 23,Wiley VCH, Weinheim

Zhonghua J, Kazuo K, Tamotsu N (1998) Major Triterpenoid Saponins from *Saponaria officinalis*. J Nat Prod 61(11):1368–1373

Zhonghua J, Kazuo K, Tamotsu N (1999) Saponarioside C, the first A-D-galactose containing triterpenoid saponin, and five related compounds from Saponaria officinalis. J Nat Prod 62(3):449–453

Chapter 4
Hair Conditioner

Abstract Nowadays the use of hair products to modify the look has become unavoidable. Hair products like shampoos, hairstyling gels, straight iron, curling iron etc. induce damage to the hair follicle as well as the hair shaft making them brittle, dry and causing split ends. One way of treating this is to stop the use of these harmful hair products or another way is to restore hair shine and softness by replenishing the lost/damaged sebum and keratin from the hair. This can be done with the use of conditioners. In this chapter few natural sources which can be used as hair conditioners along with their chemistry are described.

Keywords Banana · Avocado · Eggs · Yogurt

4.1 Introduction

Shampoos help to remove the dirt and other undesirable residues from the hair and scalp along with sebum, which is a natural hair conditioner and has to be replenished. Conditioners are divided into 5 main groups: Polymers, oils, waxes, hydrolyzed amino acids and cationic molecules. Silicones are very widely used as conditioners (Gavazzoni Dias 2015).

Functions of the conditioners are:

1. Improve compatibility
2. Minimize the hair natural lipid outer layer
3. Restore hydrophobicity
4. Seal the cuticle
5. Avoid or minimize frizz, friction: Neutralize the negative charge
6. Enhance shine, smoothness and manageability.

Ingredients of conditioners (D'Souza and Rathi 2015):

1. Cationic surfactants
2. Polymers

© The Author(s) 2016
K. Barve and A. Dighe, *The Chemistry and Applications of Sustainable Natural Hair Products*, SpringerBriefs in Green Chemistry for Sustainability, DOI 10.1007/978-3-319-29419-3_4

3. Bodying agents and thickeners
4. Emollients/oily compounds
5. Auxiliary emulsifiers

Although, there are no serious reactions reported to the use of conditioners, it is a general perspective that anything natural is much safer than synthetic ingredients. To increase public compliance and acceptability, conditioners from natural sources are also being looked into. Some of these are listed below.

4.2   *Amaranthus Spinosus*

It is commonly known as the spiny, prickly or thorny amaranth, native to tropical America.

Chemical composition

It contains a new coumaroyl flavone glycoside called spinoside, xylofura-nosyl uracil, hydroxycinnamates, quercetin and kaempferol glycoside, betalains, betaxanthin, betacyanin, phenolic compounds, amaranthine and isoamaranthine. It also contains betasitosterol glycoside, campesterol. The chemical analysis of leaves and stem gave hentriacontane and aspinasterol, linoleic acid, rutin and betacarotene as prime phytoconstituents. The bark shows the presence of betalains-amaranthine and isoamaranthine, hydroxycinnamates, quercetin and kaempferol glycosides (Stintzing et al. 2004; Mathur et al. 2010).

Benefits for hair

Amaranth leaves are a source of manganese, iron, copper, calcium, magnesium, potassium and phosphorus necessary for maintaining proper mineral balance in the body. Besides regular consumption, applying juice from the leaves prevent brittle hair falling. This also retards the onset of premature greying.

4.3   Egg

Chemical constituents

It contains saturated, polysaturated and monosaturated fats. It also contains cholesterol, sodium, potassium, as well as rich amounts of proteins. It also contains vitamin A, vitamin B-6 and B-12, vitamin C, vitamin D, iron and magnesium.

Benefits to hair

Due to higher protein content, hair growth may be enhanced with the use of eggs as conditioner. These proteins fill in the weakened spots along the hair strand and thereby help to rebuild the hair. This imparts strength to hair. Formulations containing egg lecithin are available in market as hair conditioners (Mueller et al. 1991). A patent dated 1969 had made use of egg proteins (ovalbumin, conalbumin) to prepare an herbal conditioner (John 1969). Following this there are many patents filed that makes use of egg white (egg proteins) as a conditioning agent.

4.4 Honey

Many hymenopteran insects produce honey, but the one produced by honey bees, belonging to the genus *Apis*, is always of the highest quality. The nectar from the flowers is converted to honey by the process of regurgitation by the honey bees and it is stored in the beehive. Honey gets its sweetness from the monosaccharides fructose and glucose (Anonymous 2015).

Chemical composition
Honey is a complex mixture of carbohydrates, proteins, amino acids, the most abundant being proline, vitamins, minerals, antioxidants and other compounds. Essentially, natural honey is a sticky and viscous solution consisting of carbohydrate mainly glucose and fructose. It also contains water, protein, ash and minor quantities of amino acids, vitamins and phenolic antioxidants It contains a number of enzymes, including invertase, glucose oxidase, catalase, and acid phosphorylase. It contains phenolic acids such as caffeic acid, isoferrulic acid, Gallic acid, p-coumaric acid, 4-Hydrobenzoic acid, syringin acid, and flavonoids such as quercetin, luteolin, 8-methoxykaempferol, pinocembrin, isorhamnetin, kaempferol, pinobanksin. Some other acids found in honey include butanoic, citric formic, gluconic, lactic, malic, pyroglutamic and succinic acid (Buba et al. 2013; Alvarez-Suarez et al. 2014; Ediriweera and Premarathna 2012).

Benefits for hair (Madnani and Khan 2013)
Due to the high sugar content, honey prevents the loss of moisture and thereby acts as a humectant. The conditioning effect of honey is due to the high moisture content which helps to keep the cuticle soft. It can be used along with coconut or olive oil as well as with lemon for better conditioning effect. It has been used in many cosmetic formulations in a concentration ranging from 1 to 10 %. It exerts an emollient, soothing, humectant and hair conditioning effect and is an ingredient in lip ointments, cleansing milks, hydrating creams, after sun, tonic lotions, shampoos, and conditioners (Burlando and Cornara 2013). It can also be used for the treatment of dandruff associated with seborrheic dermatitis and associated hair loss (Al-Waili 2001). It imparts hair abundance and lubrication when incorporated in a shampoo (Borchorst 1978). It is incorporated in shampoos to impart lustre to hair (Mabratu 2003). Honey has been described in a patent to be used for hair loss, restoring original hair colour and in other hair and scalp disorders (Mahmoud and Mousa 2001).

4.5 *Malus Domestica*

Apple cider vinegar, otherwise known as cider vinegar or ACV, is a type of vinegar made from apple (*Malus domestica*), has a pale to medium amber color.

Chemical constituents
Organic ACV contains up to 5 % of acetic acid, apart from traces of malic, citric and lactic acid. It also contains Vitamin C, Vitamin B1, Vitamin B2, Iron, Pectin, Magnesium and Phosphorous.

Benefits to hair
Apple cider vinegar (ACV) has long been used as a natural hair care product. Due to the acetic acid content it has an antibacterial effect. It has also been used as a cleansing and conditioning agent. It is mainly incorporated in rinse formulations wherein it helps to remove the build-up of hair cosmetics and balances the pH of hair. It can treat clogged hair follicles due to bacterial infection that creates crusty flakes on the scalp (which can result in hair loss) and also stimulates better circulation to the hair follicles, which strengthens the hair roots and promotes a healthy hair growth. It can be used for extra conditioning after shampoo and can be used once or twice a week.

4.6 *Musa* Genus

These are fruits of the genus Musa, also called as plaintains. The fruits has a soft flesh covered with a rind. The flesh mainly contains starch. Two very commonly used species to get the fruits are *Musa acuminata* and *Musa balbisiana*.

4.6.1 *Chemical Composition*

Ripe banana has moisture content—61.3 %, protein content—3.15 %, ash content—6 %, fat content—1.2 %, crude fibre—1.11 %, sugar content—12.8 %, carbohydrate—27.24 %, and total solid—38.7 g/100 g (Egbebi and Bademosi 2012). It is an excellent source of potassium and can provide 23 % of daily potassium intake. Ripe pulp contains amino acids like alanine aspartic acid, cysteine, glycine, glutamine acid, histidine, isoleucine, leucine, lysine, serine, threonine, tyrosine and valine. It also contains phosphorous, calcium, vitamin A, B6, D, vitamin C, vitamin B_1 and vitamin B_2 (Sharaf et al. 1979).

Benefits for hair
Bananas are rich in potassium that strengthen hair, minimize breakage by restoring hair's natural elasticity, and facilitates repair of damaged hair. It also acts as a moisturizer because it contains natural oils and about 75 % water. It promotes hair

recovery and rehabilitation due to high content of minerals and other nutrients. Banana mask is perfectly suited for dry, dyed hair and for hair after the perm. It improves manageability and shine of hair, at the same time moisturizes the scalp and helps to prevent and control dandruff (Sampath Kumar et al. 2012).

4.7 *Persea americana* (Avocado)

It is commonly known as avocado or alligator pear, belongs to the family Lauraceae, indigenous to Mexico and Central America. The fruits containing a single seed are used for its conditioning effect.

Chemical composition
Saponins, tannins, flavonoids, alkaloids, phenols and steroids are found to be present in leaf, fruit as well as in seeds. Fruit and seed are rich sources of protein, fat, moisture content and fiber. It is also a good source of minerals like sodium, magnesium, phosphorous, zinc, iron and copper (Arukwe et al. 2012; Yasir et al. 2010).

Benefits for hair
A dry scalp often causes itching and other hair problems. The avocado fruit oil goes deep into the epidermis layer and helps in keeping the scalp moisturized. The oil stimulates hair follicles, thereby improving the blood circulation in the scalp. The oil detangles the hair without leaving it greasy.

4.8 Protein Hydrolysates

The protein hydrolysates mainly the keratin hydrolysates are obtained either by enzymatic or chemical hydrolysis and other hydrothermal methods from hair, feathers, nails, wool, horns and hoofs. These hydrolysates containing amino acids may also be obtained from plants such as soya, corn and wheat. However, the mechanical and protective properties imparted by keratin hydrolysates are unique as amino acids cannot restore the damaged molecular structure of hair. These hydrolysates are positively charged due to presence of amino acids, the negative charge of hair attracts them. This neutralizes the electrical charges eliminating the friction and providing a smoothing and conditioning effect.

4.9 Yogurt

Yogurt is a food produced by bacterial fermentation of milk. Fermentation of lactose in the milk produces lactic acid. This lactic acid in turn lowers the pH of the milk and forms yogurt due to some effect on the milk proteins. *Lactobacillus delbrueckii* subsp. bulgaricus and *Streptococcus thermophiles* bacteria are commonly used to prepare dairy yogurt. But, other lactobacilli and bifido bacteria may sometimes be added during or after culturing yogurt. Some countries require yogurt to contain a certain amount of colony-forming units of microorganisms.

Benefits for hair
Yogurt is full of protein; it offers nourishment that our hair needs to grow strong and healthy. It also has moisturizing properties which make it a natural conditioner.

Some natural and exotic butters may also be used as hair conditioners. Few have been mentioned below.

4.10   Cupuacu Butter

This butter is obtained from the seeds of *Theobroma grandiflorum*, which is related to cacao. It is commonly found in the Amazon region. The fruits are dried and the butter is extracted by pressing the fruits followed by purification of the butter by decanting and filtration.

Chemical constituents
The butter contains triglycerides of saturated and unsaturated fatty acids namely stearic, oleic, palmitic and arachidic acid (de Cohen and de Jackix 2009).

Benefits for hair
It has excellent emollient properties and can be used as a styling gel. It helps to prevent the loss of moisture and improves the mechanical strength of hair as it penetrates into the hair cortex at the same time forms a film on the hair shaft (Nogueira et al. 2008).

4.11   Mango Butter

It is the oil extracted from the fruit of *Magnifera indica*, family *Anacardiaceae*. The oil is semi solid at room temperature and is obtained by solvent extraction followed by acetone fractionation of the dried kernels of mango fruit (Rukmini et al. 1984).

Chemical composition
It contains almost equal quantities of oleic and stearic acid, palmitic acid along with some minor quantity of linoleic, arachidic, behenic, lignoceric, and linolenic acids (Solís-Fuente and Durán-de-Bazúa 2004). It also contains triglycerides and phospholipids.

Benefits for hair
It may be used to prevent the loss of moisture from hair (Oresajo and Pilla 2007). It makes the hair soft, restores the damaged cuticle and impart shine to it. Mango

butter may be used in the form of hot oil treatment. It is better suited for oily hair as it contains tannins which may make the hair dry.

4.12   Shea Butter

It was formerly known as *Butyrospermum parkii* and now known as *Vitellaria paradoxa*. It is commonly known as shea tree, of the Sapotaceae family indigenous to Africa. It is the only species in genus *Vitellaria*. The shea butter is extracted from seeds of the fruits.

Chemical composition
Shea butter is composed of different types of fatty acids like linoleic acid, linolenic acid oleic acid, palmitic acid and stearic acid. The triterpenes esters found in shea butter include α-amyrin cinnamate, butyrospermol cinnamate, α-amyrin acetate, lupeol cinnamate, β-amyrin cinnamate, lupeol acetate, butyrospermol acetate and β-amyrin acetate (Akihisa et al. 2011). It also shows the presence of tocopherol (Maranz and Wiesman 2004).

Benefits for hair
The use of shea butter helps in restoring the scalp and the hair follicles. It promotes the proliferation of follicular cells and promotes accelerated growth of healthy hair. It also provides a protective coating on the hair shaft thereby conditioning the hair.

References

Akihisa T, Kojima N, Katoh N et al (2011) Triacylglycerol and triterpene ester composition of shea nuts from seven African countries. J Oleo Sci 60(8):385–391

Alvarez-Suarez JM, Gasparrini M, Forbes-Hernández TY et al (2014) The composition and biological activity of honey: a focus on Manuka honey. Foods 3(3):420–432

Al-Waili NS (2001) Therapeutic and prophylactic effects of crude honey on chronic seborrheic dermatitis and dandruff. Eur J Med Res 6(7):306–308

Anonymous (2015) Carbohydrates and Sweetness of Honey. National Honey Board http://faculty. ksu.edu.sa/sharif/honey/Carbohydrates%20and%20the%20Sweetness%20of%20Honey.pdf. Accessed 27 Oct 2015

Arukwe U, Amadi BA, Duru MKC et al (2012) Chemical composition of *Persea Americana* leaf, fruit and seed. IJRRAS 11(2):346–349

Borchorst B (1978) Method of treating hair with a shampoo containing honey. US 4070452 A, 24 Jan 1978

Buba F, Gidado A, Shugaba A (2013) Analysis of biochemical composition of honey samples from North-East Nigeria. Biochem Anal Biochem. doi:10.4172/2161-1009.1000139

Burlando B, Cornara LJ (2013) Honey in dermatology and skin care: a review. Cosmet Dermatol 12(4):306–313

de Cohen KO, de Jackix MNH (2009) Chemical and physical characteristics of cupuaçu's fat and cocoa's butter. Documentos—Embrapa Cerrados 269:22

D'Souza P, Rathi SK (2015) Shampoo and conditioners: what a dermatologist should know? Indian J Dermatol 60(3):248–254

Ediriweera ER, Premarathna NY (2012) Medicinal and cosmetic uses of bee's honey—a review. Ayu 33(2):178–182

Egbebi AO, Bademosi TA (2012) Chemical compositions of ripe and unripe banana and plaintain. Int J Trop Med Public Health 1(1):1–5

Gavazzoni Dias MF (2015) Hair cosmetics: an overview. Int J Trichology 7(1):2–15

John SH (1969) Method of preparing a water-soluble protein lotion. US 3483008 A, 9 Dec 1969

Mabratu M (2003) All natural hair relaxer and conditioner. US 6537564 B1, 25 Mar 2003

Madnani N, Khan K (2013) Hair cosmetics. Indian J Dermatol Venerology Leprol 79(5):654–667

Mahmoud A, Mousa MA (2001) Honey preparations. US 6171604 B1, 9 Jan 2001

Maranz S, Wiesman Z (2004) Influence of climate on the tocopherol content of shea butter. J Agric Food Chem 52(10):2934–2937

Mathur J, Khatri P, Samanta KC et al (2010) Pharmacognostic and preliminary phytochemical investigations of *Amaranthus spinosus* (*Linn.*) Leaves. Int J Pharm Pharm Sci 2(4):121–124

Mueller R, Hoeffkes H, Seidel K et al (1991) Hair treatment compositions containing natural ingredients. US 5002761 A, 26 Mar 1991

Nogueira ACS, Haake HM, Jorge H et al (2008) Performance of Cupuassu products in hair care applications. http://www.skin-care-forum.basf.com/en/articles/hair/performance-of-cupuassu-products-in-hair-care-applications/2008/07/11?id=62404b9d-11fe-48ed-8a2c-50e05123a705&mode=Detail. Accessed on 1 Dec 2015

Oresajo C, Pilla S (2007) Black skin cosmetics: specific skin and hair problems of african americans and cosmetic approaches for their treatment. In: Berardesca E, Leveque JL, Maibach HI (eds) Dermatology: clinical & basic sciences, vol 8. Ethnic skin and hair. Informa Healthcare, USA

Rukmini C, Vijayaraghavan M (1984) Nutritional and toxicological evaluation of mango kernel oil. J Am Oil Chemists' Soc 61(4):789–792

Sampath Kumar KP, Bhowmik D, Duraivel S et al (2012) Traditional and medicinal uses of Banana. J Pharm Phytochem 1(3):51–63

Sharaf A, Ola Sharaf A, Hegazi SM et al (1979) Chemical and biological studies on Banana fruit. Zeitschrift für Ernährungswissenschaft 18(1):8–15

Solis-Fuentes JA, Durán-de-Bazúa MC (2004) Mango seed uses: thermal behaviour of mango seed almond fat and its mixtures with cocoa butter. Bioresour Technol 92(1):71–78

Stintzing FC, Kammerer D, Schieber A et al (2004) Betacyanins and phenolic compounds from *Amaranthus spinosus* L. and *Boerhavia erecta* L. Z Naturforsch C 59(1–2):1–8

Yasir M, Das S, Kharya MD (2010) The phytochemical and pharmacological profile of *Persea americana Mill.* Pharmacogn Rev 4(7):77–84

Chapter 5
Hair Colours/Dyes

Abstract Hair dyes have been used since ages to modify hair color and hence ones looks. It is also used to hide the greys as it comforts an individual. This chapter provides information on the chemistry of plant drugs which could be used as hair dyes or as hair colors and the benefits offered by the same if used in hair cosmetics.

Keywords Hair colors · Alkanna · Henna · Indigo · Walnut

5.1 Introduction

Since ancient times people have been using hair colors. Loss of color from hair may be due to various reasons like genetic influence, effect of environmental factors etc. Hair may be coloured to enhance the appearance or as a fashion statement. Depending on the penetration of hair dyes, they can be classified as Demi-permanent and permanent dyes.

Permanent hair colors need to reach the cortex, which is achieved by increasing the pH using either ammonia or triethanolamine. The natural cuticle lipid, 18-metil eicosanoic acid is removed with the use of any type of dye which induces damage to the hair shaft. The pigments used are dyes like para-phenylenediamine, para-toluenediamine and para-aminophenol with hyrodgen peroxide. Thus permanent dyeing involves oxidation reaction to allow the pigment to reach the cortex.

Demi permanent dyes are gentler on hair due to absence of strong alkali. They do not reach the cortex and remain on the cuticle surface and hence are also easily washed after about 10–15 shampoos. The major ingredients of these dyes include hydrogen peroxide, resorcinol and para-dyes. The may also be used to add a shine and to provide a vibrant colour to the natural hair.

The use of these dyes causes contact dermatitis. Although no conclusive evidence exists, there is some linkage between hair dyes and cancer hazard.

Hence there is a need for natural pigments that can function as hair dyes/colours. Following is a list of few herbs that can be used as hair colours.

© The Author(s) 2016   45
K. Barve and A. Dighe, *The Chemistry and Applications of Sustainable Natural Hair Products*, SpringerBriefs in Green Chemistry for Sustainability, DOI 10.1007/978-3-319-29419-3_5

5.2   *Alkanna tinctoria* (**Ratanjot**)

It is also known as alkanet, belonging to the family Boraginaceae. The root yields a red dye and is indigenous to the Mediterranean. The plant has a dark red root is blackish red externally but blue-red with a white core internally. The dye is in use in the Mediterranean region since antiquity (Khan et al. 2015).

Chemical constituents
The leaves are reported to contain alkaloids, carbohydrate, flavonoids, proteins, resins, saponins, tannins, phenolics, triterpenoides and steroids. The root contains a mixture of red pigments consisting of mainly of fat-soluble naphthazarin, a napthaquinone derivative and other components such as alkannin and related esters. Beta-dimethylacrylic acid, beta-acetoxy-isovaleric acid, isovaleric acid, and angelic acid are the acids found the alkannin esters, isolated from the root.

Benefits for hairs
The alkanet root when soaked in coconut oil will cause a change in colour of the oil and this can be applied to hair for hair color purpose. It has been reported to be used as a hair dye since the Medieval ages (Cavallo et al. 2008).

5.3   *Cassia obovata/Senna italica*

It is commonly known as neutral henna, blonde henna or colorless henna giving a yellow tinge to hair. It belongs to the family Leguminosae.

Chemical constituents
It contains anthraquinone glycosides like sennosides A and B, C and D. Rhein, aloe-emodin and chrysophanol are the free anthroquinones reported in the plant. It also has polysaccharides and mucilage consisting of galactose, arabinose, rhamnose, and galacturonic acid. Flavonols like isorhamnetin and kaempferol and glycosides like 6-hydroxymusizin and tinnevellin are also found in the leaves (Bisset 1994; Duke 1985).

Benefits for hairs
The cysteine group of keratin in the hair absorb *C. obovata* components, thus it may protect from the UV rays and act as a sunscreen (Forestier 1981). When applied to, in an acidic juice to hair it works to prevent premature greying of hair. It may not be used as a colour on its own but in combination with other natural hair colours it produces different effect. In combination with rhubarb root it enhances the golden tone, with pure henna it enhances the strawberry tone and with catechu it enhances the dark tone. Along with that it provides a conditioning effect, promotes hair growth and also helps to cure dandruff.

5.4 *Crocus sativus*

Saffron is a spice derived from the flower of *Crocus sativus*, commonly known as the "saffron crocus". Saffron crocus grows to 20–30 cm (8–12 in.) and bears up to four flowers, each with three vivid crimson stigmas, which are the distal end of a carpel.

Chemical constituents
Saffron contains carotenoid pigments: crocin responsible for the colour, picrocorocin responsible for the flavour and a volatile component: safranal responsible for the earthy fragrance. Safranal is formed on breakdown of picrocrocin due to the effect of enzymes or heat. It also contains other carotenoids lycopene, zeaxanthine, and both alpha- and beta-carotenes.

Benefits for hairs
When a mixture of water boiled with saffron is applied to hair and later washed with a shampoo, it leaves grey hair turn dark golden brown which complements your already dark hair. Saffron was used by the Romans to give a blonde tint (Roia 1966).

5.5 *Eclipta prostrate/Eclipta alba* (Maka, Bhringraj)

In India it commonly known as bhringraj or bhangra and belongs to the Asteraceae family. The plant has traditional uses in Ayurveda as the one supporting growth of hair and colour. The white-flowered *E. alba* is called white bhangra and the yellow-flowered *W. calendulacea* is called yellow bhangra (Mithun and Shashidhara 2011).

Chemical constituents
It contains the coumestan derivatives, which have coumarin as their base structure like wedelolactone, demethylwedelolactone and demethylwedelolactone-7-glucoside abundant in the leaves. It also shows the presence of glycosides, triterpenoids, alkaloids and flavonoids. The roots contain long chained hydrocarbon alcohols like hentriacontanol and heptacosanol and polyacetylene substituted thiophenes. The aerial part is reported to contain a steroids, triterpenoids and coumestans (Mithun and Shashidhara 2011).

Benefits for hairs
The herb has a traditional use in the treatment of hair loss. It is also used to prepare a black dye for colouring hair. These extracts are known to grow hair faster than minoxidil (Roy et al. 2008). Oil containing bhringraj juice promotes hair growth and prevents premature greying. It is further reported to be used in the treatment of alopecia by stimulating follicular keratinocyte proliferation (Begum et al. 2015; Datta et al. 2009).

5.6 *Indigofera tinctoria* (Indigo)

It belongs to the family Fabaceae distributed in the tropical and subtropical regions of the world. *Indigofera tinctoria* and *Indigofera suffruticosa* are the species used to produce the dye indigo.

Chemical constituents
Dye is obtained from the processing of the plant's leaves. These are soaked in water and fermented in order to convert the glycoside indican naturally present in the plant to the blue dye indigotin. Indican was obtained from the processing of the plant's leaves, which contain as much as 0.2–0.8 % of this compound.

Benefits for hair
The application of powdered plant material in hot water, at least 20–40 min before washing for every 4–6 weeks imparts colour to the hair. Natural indigo may be present up to 0.17 % in final hair dye formulation. Indigo powder is considered as non-irritating to the skin after performing the patch test on animals but this is considered as irritant to eyes at the same time. Use of indigo can give shades ranging from light red brown to black, also imparts a blue tone. It can be mixed with henna for application, but doesn't stay for a long time. Indigo carmine can be used with other natural dyes for natural dyeing and for colouring hair (Komboonchoo and Bechtold 2009). A combination of henna and indigo produces a suitable brown colour. Repeated application increases the intensity of the colour. Addition of ferric chloride may enhance the colour retaining property (Rao and Shayeda 2008).

5.7 *Juglans nigra* (Black Walnut)

It is commonly known as black walnut, belongs to the Juglandaceae and is indigenous to North America. The tree has a high value for its wood. The nuts were used by the local people to dye their hair.

Chemical constituents
It contains juglone a naphtoquinone derivative found in the walnut hull, leafs, bark and even roots, as its active ingredient. The other constitutes include tannins, plumbagin and iodine.

Benefits for hairs
Black walnut drupes (the outer part of the walnut; the hull) contain juglone, plumbagin, and tannin (like in teas). This allows it to dye pretty much anything it touches. It can be used as a wood stain, ink, hair dye, clothing dye, etc. (Gupta and Gulrajani 1993). Application of juglone imparted a darker brown colour to the hair (Youn-Sook and Soo-Hee 2006).

5.8   *Lawsonia inermis* (Henna)

It is commonly known as henna belonging to the family Lythrceae, indigenous to Africa, Asia and Australasia. The use of henna has been intricately woven in the ancient and modern cultures of Asia used for dying skin, hair, nails and even fabrics (Bailey and Bailey 1976).

Chemical constituents
It contains Lawsone, a napthoquinone derivative as an active constituent. B-ionine is the one responsible for the pungent odour of the flowers (Semwal et al. 2014). It also contains volatile components like linalool, α terpineol, 1,3-indandione, eugenol etc. (Kebedekidanemariam et al. 2013).

Benefits for hair
Henna stains the skin reddish brown and is known as red henna, the same when mixed with p-diphenylenediamine is termed as black henna. The keratin in hair binds with lawsone, this penetrates the hair shaft but not permanently. Henna when incorporated in different dye formulations may provide different shades to hair (Singh et al. 2015). The colouring principle of henna which is lawsone, is reported to act as potent oxidant of G6PD deficient cells. A word of caution may be used in context to henna, as it is reported to cause life threatening hemolysis in children deficient of G6PD (Gavazzoni Dias 2015).

References

Bailey LH, Bailey EZ (1976) Hortus third: a concise dictionary of plants cultivated in the United States and Canada. Macmillan, New York
Begum S, Lee MR, Gu LJ et al (2015) Exogenous stimulation with *Eclipta alba* promotes hair matrix keratinocyte proliferation and downregulates TGF-β1 expression in nude mice. Int J Mol Med 35(2):496–502
Bisset NG (1994) Herbal drugs and phytopharmaceuticals. CRC Press Inc., Stuttgart
Cavallo P, Proto MC, Patruno C et al (2008) The first cosmetic treatise of history. A female point of view. Int J Cosmet Sci 30(2):79–86
Datta K, Singh AT, Mukherjee A et al (2009) *Eclipta alba* extract with potential for hair growth promoting activity. J Ethnopharmacol 124(3):450–456
Duke J (1985) CRC handbook of medicinal herbs. CRC Press Inc., Boca Raton
Forestier JP (1981) A cosmetic senna, *Cassia obovata*: 'neutral henna'. Int J Cosmet Sci 3(5): 211–226
Gavazzoni Dias MF (2015) Hair cosmetics: an overview. Int J Trichol 7(1):2–15
Gupta DB, Gulrajani ML (1993) Studies on dyeing with natural dye Juglone (5-hydroxy-1, 4-naphthoquinone). Indian J Fibre Text Res 18(4):202–206
Kebedekidanemariam T, Tesema TK, Asressu KH et al (2013) Chemical investigation of *Lawsonia inermis* L. leaves from afar region, Ethiopia. Orient J Chem 29(3):1129–1134
Khan UA, Rahman H, Qasim M et al (2015) *Alkanna tinctoria* leaves extracts: a prospective remedy against multidrug resistant human pathogenic bacteria. Complement Altern Med 15:127

Komboonchoo S, Bechtold T (2009) Natural dyeing of wool and hair with indigo carmine (C.I. Natural Blue 2), a renewable resource based blue dye. J Clean Prod 17(16):1487–1493

Mithun NM, Shashidhara S (2011) *Eçlipta alba* (L.) a review on its phytochemical and pharmacological profile. Pharmacologyonline 1:345–357

Rao YM, Shayeda, Sujataha P (2008) Formulation and evaluation of commonly used natural hair colorants. NPR 7(1):45–48

Roia FC (1966) The use of plants in hair and scalp preparations. Econ Bot 20(1):17–30

Roy RK, Thakur M, Dixit VK (2008) Hair growth promoting activity of *Eclipta alba* in male albino rats. Arch Dermatol Res 300(7):357–364

Semwal RB, Semwal DK, Combrinck S et al (2014) *Lawsonia inermis* L. (henna): ethnobotanical, phytochemical and pharmacological aspects. J Ethnopharmacol 155(1):80–103

Singh V, Ali M, Upadhyay S (2015) Study of colouring effect of herbal hair formulations on graying hair. Pharmacognosy Res 7(3):259–262

Youn-Sook S, Soo-Hee L (2006) Natural dyeing of hair using juglone. J Korean Soc Clothing Text 30(12):1708–1713